St. Louis Community College

Library

5801 Wilson Avenue
St. Louis, Missouri 63110

Fred L. Whipple

The Mystery of Comets

Assisted by Daniel W. E. Green

Smithsonian Institution Press

Washington, D.C.

1985

Library of Congress Cataloging in Publication Data
Whipple, Fred Lawrence, 1906-
The mystery of comets.
Supt. of Docs. no.: SI 1.2:C7313
1. Comets—Popular works. I. Title.
QB721.4.W47 1985   523.6   85-8343
ISBN 0-87474-968-9
87474-971-9 (pbk.)
Includes index.

Frontispiece: A typical bright comet as it might
appear to the naked eye. This 1970 photograph by
S. L. Larson pictures comet Bennett, 1970 II, in the
morning sky over the southwestern United States.

To Babette

without whom this book

would not have been written

Contents

Foreword

In the past twenty-five years, the study of the planets and other bodies in the Solar System has expanded from the realm of astronomers into that of geologists, meteorologists, chemists, and physicists. These scientists have been able to gather a wealth of data by means of numerous manned and unmanned spacecraft launched during this period. Although the voyages of these spacecraft provided the basis for further exploration of the Solar System, in some cases they came so fast that there was little time to assimilate all the information returned. With the hiatus in planetary missions during the mid-1980s, however, scientists have been able to concentrate their attention on analyzing the data collected thus far. Upon completion of the National Air and Space Museum's *Exploring the Planets* gallery, Robert W. Wolfe, curator of the exhibit, suggested publication of a series of books that would explain the status of our investigations of other bodies in the Solar System and convey the excitement of our new discoveries. That is but one objective of the Smithsonian Library of the Solar System.

The next decade will see a rebirth of exploration of the planets and other bodies in the Solar System, as plans are now being finalized for the Galileo Mission to Jupiter, the Mars Observer, the Venus Radar Mapper, and the Comet Rendezvous/Asteroid Flyby Mission. We hope that this series will also help a wide audience to understand why these missions are needed and how their new discoveries will change our concepts of the evolution of our neighbors in space, and the evolution of our Earth itself.

This series would not be possible without the work of the dedicated scientists and engineers who plan, develop, and carry out the exploration of the Solar System. We are indebted to the National Aeronautics and Space Administration for supporting these goals and the scientific research that provides the basis for future missions. We are especially grateful to Walter J. Boyne, Director of the

National Air and Space Museum, and to Felix C. Lowe, Director of the Smithsonian Institution Press, for their support of this series.

Ted A. Maxwell
Center for Earth and Planetary Studies
National Air and Space Museum

Preface

Fascination with comets stretches back into the dim past, certainly beyond the sporadic records of the ancient Chaldeans, Chinese, Greeks, and Romans. Throughout these many centuries, thoughtful naturalists repeatedly attempted to explain, within the current or their own cosmological framework, the true nature of those rare and spectacularly beautiful visitors to our skies—the comets. In his own way, each of these thinkers was trying to replace superstition and incorrect ideas with a rational philosophical explanation for a mystifying celestial phenomenon. Of necessity, lasting progress had to wait until after the Middle Ages, when the scientific method emerged and permitted mankind gradually to build up a consistent, stable body of knowledge about the Solar System and the Universe. Even so, only recently have scientists unraveled parts of the cometary mystery and put together a coherent picture of what comets are and why they behave the way they do.

This book tells the story, step by step, from ancient times to the present era, of our progress in understanding comets. The book divides into two sections: The first fourteen chapters present clues and theories developed before the twentieth century; the remaining chapters provide up-to-date information regarding the nature of comets, their origin, and their possible relation to life on Earth. Comets appear to be the most primitive bodies in the Solar System and may even have played an important role in making life on Earth possible. Later chapters center on these issues, on space missions to comets (particularly to the famous comet Halley), and on the general problems and significance of comets, including the hazards they may represent. The story should be intelligible to readers with little or no background in science, even though they may find themselves intellectually stretched by some sections. Scientifically sophisticated readers, on the other hand, will review familiar territory at first, then will encounter more challenging material as they proceed.

Science as a whole has developed from somewhat random observations of apparently unrelated phenomena, interlaced with many speculations, to an organized system of observations and experiments, soundly linked together by mathematical theory. Our understanding of comets has grown in parallel. The story of comets cannot be fully appreciated unless it is related to the overall evolution of science, in particular to that of physics and Solar System astronomy. Nor can we ignore the enormous role of technology in helping us to understand comets—technology is both the offspring of science and the foundation for its continuing progress. Hence I will present the growth of cometary understanding as an integral part of the fantastic growth of science and technology.

I am particularly indebted to Daniel W. E. Green who greatly reduced my tiresome effort by critically reading and improving the text, helping research many concepts in the book, and scouting for illustrations. I am also grateful to Donald K. Yeomans for helping me to obtain rare illustrations. Brian Marsden critically read several chapters. Many others who provided much appreciated assistance are identified in the figure captions. Linda McKnight drew or redrew almost all of the diagrams.
The wise guidance of I. Bernard Cohen increased the accuracy of the Newton-Halley discussion. David Pingree and Anthony F. Aveni helped me greatly with the opening chapters. My wife, Babette, read all of the early texts and prodded me toward the path of precise and intelligible expression.

Chapter 1 The Fear of Comets

Why should comets—those graceful, sometimes majestic, creatures of the sky—frighten people? They move very slowly, without startling changes in shape or aspect. They make no sounds and emit no dazzling flashes of light. In short, they do nothing that seems to me to be threatening. Yet comets have terrified people as long as there have been people to terrify. In ancient Greece, for example, comets had already gained top billing as malicious and dangerous entities. Around 900 B.C., Homer used the image of a comet to inspire fear in his audience when describing Achilles' helmet:

Like the red star that from his flaming hair
Shakes down disease, pestilence and war.

The "flaming hair" refers, of course, to the great tails or "tresses" that stream across the sky from the heads of bright comets. The word "comet" itself comes from the Greek word *kometes*, meaning "hairy one." The Greeks credit the Egyptians with the idea that the tail of a spectacular comet resembles a woman's long hair. But elsewhere in ancient descriptions, the tail is an axe, a scimitar, a sword, or a dagger. Clearly we are dealing with a twisted version of the old saying, "beauty lies in the eyes of the beholder." The fearsome look of comets lies deep in the mind of the viewer. And in ancient times, comets could be seen by everyone. Until very recently, all those who had the courage to brave the night could see a bright comet. This was true for city folk as well as for shepherds or country folk. Only in the last century have we fouled the night skies of our cities with lights and smog to such an extent that bright comets are practically invisible to city dwellers today.

Comets appear without warning. They appear anywhere in the sky; some last for days, others for weeks or months. They move sedately in any old direction among the stars, easily interpreted as an indication of wilful intent, undoubtedly malicious. At least the "wandering"

1

Excerpt from work by Kitab al-Mughni, astrologer at Baghdad in the ninth century. The planetary classification follows the Greek. Top (left to right): bearded, Jupiter; lamp, Mars; spear, Mercury; horseman, Venus. *Center row:* giant, Sun; "the girl," Moon; kettle, Saturn. *Bottom row:* "circular," lamp, skewer, jar, javelin. (Interpretation by David Pingree; courtesy Princeton University Library.)

stars or *planets* stay near the Zodiac, the great circle in the sky along which the Sun makes its yearly march. The major source of the fear of comets must lie in their erratic behavior, coupled with some acquired element of human culture.

Primitive humans gained ascendancy in the world through their ability to pass on their hard-earned experience to following generations. They thereby developed the power of prediction—with respect to changes in the seasons, varying sources of food, the imminence of danger, and so on. Imagination and the power of prediction greatly affect human thinking and human life. The individual feels a sense of assurance and satisfaction when all goes as expected, but experiences instinctive fear and tension when confronted with the unexpected or the unpredictable. Once human beings became hunters and gatherers, they were exposed to many real dangers, such as wild animals, food scarcity, lightning, storms, floods, and the like. We may speculate that comets added only a little to the huge fear-budget of these people, although such apparitions in the sky must have been disturbing. Accurate prediction of the seasons was probably not yet important because the seasons could be observed well enough for the needs of early man in the arrival of fruits and berries or in the migrations and habits of hunted animals.

As human cultures advanced to the level of agriculture, the question of when to plant demanded more in the way of prediction. This development must have increased or possibly initiated the function of the astrologer, in addition to the duties of the medicine man or the tribal wise man. He now kept track of the Sun's rising and setting as it moved along the horizon with the seasons, particularly in the high latitudes where seasonal changes are so important. Some societies focused on lunar changes and others on predictions associated with the appearance of Venus or the brightest stars. Comet watching, however, was a different story. The unexpected event—such as the appearance of a bright comet—could arouse deep-seated fears of the unknown, the uncontrollable. The interpretation of these and other omens of nature then became an important duty of the local authority in such matters, and

This 1857 French cartoon depicts a fear that still exists today—that a comet may hit the Earth and cause widespread destruction.

in time became the basis of traditional folklore, mythology, and superstitions. All the uncontrollable events of nature—whether repeated or unexpected—required an explanation, however bizarre it might seem to us today. Once stated, the explanations were generally accepted, and around them developed special ceremonies, taboos, feasts, and celebrations to make people feel at ease under worrisome situations. A current example of this practice can be found in the midwinter rites celebrated when the Sun reaches its maximum southern extremes in the northern hemisphere. The hope and promise of the spring to come is a source of festivity, Christmas for us. We also carry over the ancient harvest festivals in Thanksgiving Day.

With the rise of cities, empires, and other large-scale organized societies, rulers became particularly sensitive to the omens of nature. A ruler knew that his power and even his life were precarious assets. Any means of predicting the future, particularly any means of altering it favorably, was highly welcome. In China for many centuries, the emperors symbolized the heavens; that is, they were the embodiment of supernatural powers. Thus the Chinese emperors became more interested in celestial phenomena than most of their counterparts in other great cultures. They encouraged astrologers to observe and record the heavens on a daily basis. The astrologer's duty was to report immediately all portents such as comets, meteors, northern lights, and "new" stars. Ingenious interpretation of these omens in terms of dangers to come, coupled with sagacious advice as to effective countermeasures, was undoubtedly highly rewarded and thus considered a form of job insurance. Meanwhile, any lack of diligence on the part of the court astrologers was severely punished. The case is told of two ancient Chinese astrologers, Hi and Ho, who failed to predict an eclipse of the Sun. The emperor had them beheaded for dereliction of duty (probably for being drunk), at least according to legend.

Chinese records have been a gold mine for comet study. They include descriptions and dates of some 600 comets, the earliest being about 2315 B.C., according to the translations of the Japanese historian Ichiro Hasegawa. The Chinese observed the paths of many "broom stars" among the stellar constellations of their sky maps. Their "zodiac," by the way, contains twenty-eight "lunar mansions" instead of the twelve "signs" adopted by the Western world. Halley's famous periodic comet can be traced back to 240 B.C. (possibly to 1615 B.C.) in the Chinese writings. Until the seventeenth century, Western astrologers were more interested in interpreting the dire forebodings of the comets than in keeping track of their movements on the sky. We owe an enormous debt to these Chinese recordkeepers because comets carry no name tags. The only way we can identify them is by their position among the stars at a given time.

Other highly developed ancient cultures have been much less helpful than the Chinese in our comet story. The Sumerians and Akkadians in the Euphrates River valley left countless baked clay, highly organized records in cuneiform, or wedge-shaped writing. David Pingree of Brown University, a leading scholar in this field, in a letter to me writes that these tablets "contain numerous omens where celestial phenomena—undoubtedly including comets—are taken as indications of terrestrial events. The oldest of these texts go back nearly four thousand

Astronomical records of the Aztecs in the postconquest Codex Telleriano-Remensis show a number of comets, citlalimpopoca, *or stars that smoke. (Courtesy Anthony Aveni,* Skywatchers of Ancient Mexico *[Austin, Tex., 1980].)*

years to the time of Hammurabi, but for that early period, the Akkadian word which means comet has not yet been securely identified. In fact, we can be certain of the existence of such a word only in cuneiform texts written in the third and second centuries B.C." Pingree has indeed found an unpublished record of a comet observed in Babylonia in November 164 B.C., almost certainly the second oldest known observation of Halley's comet. His late colleague, Abraham Sachs, made the original translation from cuneiform.

The earliest classification of comets appears in a Greek omen text from about 150 B.C., attributed falsely to the Egyptian priest Pelósiris. In this text, the different types of comets are associated with the planets: the *horseman* with Venus; the *swordsman* and the *torch-holder* with Mercury; the *long-haired* with Jupiter; the *disk-thrower* with Saturn; and the *typhoon* with Mars. Descriptions of the various types of comets and their astrological effects are also included.

Babylonian tradition accounts for a classification of comets recorded in India around A.D. 100. Here comets were divided into three categories: terrestrial, atmospheric, and celestial, the last being the "sons" of various deities, including the planets. All the types were described.

In the New World, the Spanish conquistadors destroyed almost all the extensive "libraries" of the Mayans and the Aztecs. Nevertheless, the remaining fragments and the accounts of the early Spanish chroniclers indicate that the rulers in these great cultures revered the heavens and were as much in awe of them as were the Chinese emperors. In the early centuries A.D., Mayan astronomy and timekeeping may have surpassed the ideas and practices in Europe and Asia. The description that follows relies on intensive studies of Meso-american astronomy made by Anthony F. Aveni of the University of Texas. In the cosmology of the Meso-american priest, layers existed above and below the flat Earth; each of the thirteen upper layers contained a single heavenly body or god. The Moon moved in the first layer above the Earth, then the clouds, the stars, the Sun, Venus, comets, and so on. The Aztecs called comets "stars that smoke" and depicted them as stars from which emanated stylized puffs of smoke. The chronicler Father Diego Durán reports that the powerful Aztec ruler Moctezuma actually observed a great comet himself. He then chided the astrologers, soothsayers, and diviners for carelessness in observing, interpreting, and

A great comet was said to have predicted the fall of Moctezuma's empire. (Courtesy Empresa Editorial Cuautémoc, Mexico.)

reporting the comet. Legend has it that this great comet presaged the fall of Moctezuma's empire and the final demise of that great culture. Even though records are sparse, they clearly indicate that the Aztecs and Mayans worried about comets in much the same fashion as did the peoples of the Old World.

The Chumash Indians of California developed a highly imaginative mythology of the stars and the heavens, as gleaned from intensive studies by Travis Hudson and Ernest Underhay of Santa Barbara, California. A pair of Chumash rock glyphs appears to represent two comets or two views of one comet. The Chumash seemed to regard comets as bad omens. In contrast, when the medicine man of the Wampanoog Indians in Massachusetts, Mr. Slow Turtle Peters, was recently interviewed by Bernard M. Taylor of Boston and asked about his interest in comets, he replied, "I couldn't care less." Assimilated into the melting pot of European cultures, these Indians have been detached from their old traditions, which may have included comet watching.

Returning now to prescientific attitudes toward comets in Europe, we note that astrologers had a field day when a rare, bright comet came along. Clearly, such celestial monstrosities had to portend great events. In Shakespeare's *Julius Caesar*, we find:

When beggars die there are no comets seen.
The heavens themselves blaze forth the death of Princes.

Shakespeare was no economist, of course, but he must have recognized a clear-cut case of supply and demand when he saw one. There simply are not enough comets to go around. They must be reserved for great rulers, kings, and emperors. In Roman times, almost everyone subscribed to this idea. The comet of 44 B.C. appeared *after* the assassination of Julius Caesar, however—too late to be the harbinger of his death. To maintain consistency and the symbolic importance of comets, the Romans agreed that this comet carried his soul into the ranks of the immortal gods. Typical was the belief that the comet of A.D. 336 had announced the death of the great Emperor Constantine. Dozens of such cases appear in the Dark Ages; comets "presage" the deaths of Attila, the Emperor

Rock art comet motifs by Chumash Indians of California. (Sketched by Campbell Grant. Courtesy Travis Hudson and Ernest Underhay, Crystals in the Sky, Ballena Press, 1978.)

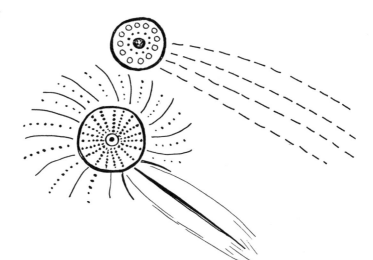

Medal struck in Germany to allay the fear of the comet of 1680. "The star threatens evil things: Only Trust!/God will make things turn to good." (Courtesy Donald K. Yeomans.)

Valentinian, a number of other emperors, and several popes.

In his classic painting of the nativity scene, the great Florentine artist Giotto depicts the star of Bethlehem as a comet. He was probably inspired by the great comet of 1301, actually Halley's comet, observed in Europe as well as in China. We now know that the only bright comet near the time of the nativity probably appeared in 12 B.C. (also Halley's comet). Because of the famous painting, the European Space Agency has given its 1986 space mission to Halley's comet the name *Giotto*.

Chapter 2 Early Theories of Comets

The best inquiring minds of the Chaldeans, Egyptians, Greeks, and Romans made serious attempts to explain comets. The eighteenth-century French astronomer and historian Alexandre Pingré grouped their ideas into three classes: (1) those that claimed comets are truly heavenly bodies; (2) those that argued they are close to the earth, and are atmospheric/meteorological phenomena; and (3) those that considered them to be optical illusions. We need not dwell on the arguments that comets are optical illusions. Few thinkers, none of them influential, held this belief.

In the fourth century B.C., Aristotle, the greatest scientist of antiquity, strongly espoused the atmospheric theory. Aristotle's cosmology was greatly influenced by the Pythagorean philosophers and mathematicians and by his older contemporary, Plato, who believed that the only immutable, thus real, perfection lies in geometry and in the distant heavens. Everything near or on the Earth is accompanied by injustice, corruption, and decay. The daily cycle is produced by the rotation of the heavens about a fixed Earth. Thus, to Aristotle, the stars appeared to be attached to a perfect, incorruptible, crystalline sphere, an invisible sphere that turns about the fixed Earth. To complete the concept, each of the five planets, along with the Sun and the Moon, has its own invisible, crystalline sphere. Aristotle ended up with fifty-six such spheres, having added enough to counteract their disturbing effects on each other. The Moon travels on the smallest sphere nearest to the Earth. The daily motion within this sphere heats the high atmosphere to produce lightning, shooting stars (meteors), aurorae or northern lights, and other atmospheric phenomena. Comets are merely earthly exhalations raised to this fiery region, ignited and carried along with the motions of the hot air that they encounter. Needless to say, no one ever attempted to use Aristotle's system for prediction.

Aristotle's dictum that comets are earthly exhalations

did not go unchallenged. A number of ancient philosophers maintained that comets are wandering heavenly
bodies like the planets, but different in appearance and
following somewhat different paths or laws of motion.
Most remarkable was the insight of Seneca, advisor to the
Roman Emperor Nero. Seneca, living some 400 years
after Aristotle, asked whether the Earth itself might be
rotating, instead of the invisible celestial spheres. Seneca
considered comets to be a type of planet and he predicted:

Some day there will arise a man who will demonstrate in what
regions of the heavens the comets take their way; why they jour-
ney so far apart from the other planets; what their size, their
nature.

His nongeocentric argument could not stem the forceful
tide of Aristotle's opinions, however. Small wonder that it
took hundreds of men and women and many hundreds of
years to answer the basic questions about comets, even in
part. The search goes on.

In Medieval Europe, the Church took little notice of
comets, possibly because they are not mentioned specifically in the Bible. A few passages may refer to comets, the
best example being from 1 Chronicles 21:16,

*The great comet of
1811. Napoleon's
"Nemesis"? (Courtesy
Donald K. Yeomans.)*

The two comets of 1472 and their ominous signs. Ursula Marvin translates the German text that accompanies the woodcut as follows: "Concerning a comet, which was called the peacock's tail or the whip. All these things occurred in the year since the birth of Christ 1472. Between about sundown and midnight a comet appeared with a long, dark, streaming tail; and just as this one had not quite disappeared, another came up at sunrise with a streaming tail like a whip. Several people said it was like a peacock's tail. This produced such a dry summer that no fruit grew and everything was very dry. Also a great plague came in all the German lands, and much war followed in the German and Italian lands; there was heard such consternation as was never heard before." (Reprinted from the Lucerne Chronicle by Diebold Schilling [1513], Folio 77; courtesy Zentralbibliothek, Lucerne.)

And David lifted up his eyes, and saw the Angel of the Lord
stand between the Earth and the Heavens, having a drawn
sword [comet?] in his hand stretched out over Jerusalem.

The story that Pope Calixtus III actually excommuni-
cated the comet of 1456 (an apparition of Halley's comet)
is a hoax. The Pope was clearly worried, however. He
ordered public prayers for deliverance from the comet
and from the enemies of Christianity. Its enemies at that
time were the Turks, who were besieging Belgrade. Apparently not only were the Pope and the public numbed by
fear of that comet, but also the two opposing armies.

For nearly two thousand years, the Western world recognized Aristotle as the fountainhead of scientific knowledge. Seneca's ideas were ignored while Aristotle's word
was absolute and final in natural philosophy. Because
comets were unpredictable and seen as confined to the
Earth's atmosphere, any serious study of them was
deemed unworthy by the philosophers and was not
encouraged by the Church. This attitude produced two
unfortunate consequences: First, Europeans recorded few
careful observations of comets before the bright comet of
1472, a disgraceful showing when compared with that of
the Chinese; second, the public had no explanation for
these frightful apparitions, except for the dire pronounce-
ments by astrologers and notable public figures.

Let us now check on the progress of comet theory by the
thirteenth century. The stirrings of new ideas are evident,
but not ideas about comets. Roger Bacon is often credited
with being the father of modern science. Although he
advocated the use of experiments and reason, he was far
from independent of Aristotle's authority and astrological
tradition. The great comet of 1264, with a tail 100 degrees
long, so impressed Bacon that even he remarked it was
"great and dreadful" and noted "the comet's appearance
was followed by [that is, caused] vast disturbances and
wars in England, Spain, Italy, and other lands, in which
many Christians were slaughtered." He further stated
that the comet moved rapidly toward the planet Mars, by
whose "force" it was "generated."

We owe a debt to the comet of 1264 because it led Aegidius of Lessines, a Dominican scholar, to write a compre-

Seven drawings of classes of comets depicted as swords with handles, by Johannis Hevelius, a student of comets in the seventeenth century.

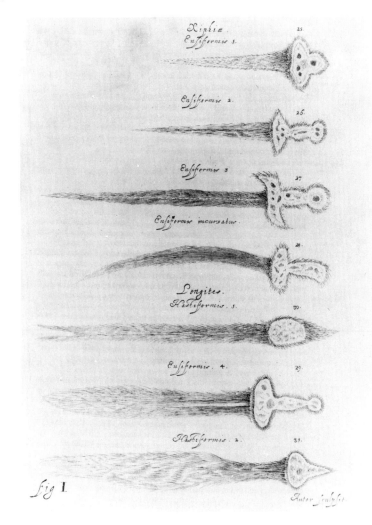

hensive treatise on comets, based on the surprisingly vast literature (containing few observations) from antiquity to the writings of his time. We thus have a fix on the best scholarly opinion about comets near the end of the thirteenth century, shortly before the Renaissance.

Basically, Aegidius accepted Aristotle's theory and dismissed Seneca's view on the grounds that comets are of too brief duration and are too changeable in appearance to be genuine heavenly bodies. He denied that comets are special creations of Providence, that is, divine signs. His reasoning on this point was in tune with the predominant religious thinking of his time: which was that divine signs are not given to everyone, but to infidels rather than the faithful. He also noted that comets seem to be natural phenomena, and can be predicted from conjunctions of planets; the last statement was based on the writings of Albumasar, whose theories proved to be incorrect. He discussed at length the problems of cometary motions, whether the tail is a fundamental part of a comet or merely a reflection, and the existence and meaning of different kinds of comets. All in all, we can see that, by the thirteenth century, little progress, if any, had been made in solving even the mystery of whether comets are atmospheric or heavenly. The unfolding story still contains only a few disconnected clues. Our scientific detectives have yet to place the Earth properly among the planets and the Sun in the Solar System.

During these many centuries the practical problems of navigation and of predicting the seasons and the positions of the Sun, Moon, and planets on the sky were solved by reliance on a system devised by Ptolemy of Alexandria about 130 B.C. Ptolemy was a superb mathematician and astronomer. From as far back as ancient Babylonia he compiled observations and studies of the Sun, Moon, and the five bright planets (Mercury, Venus, Mars, Jupiter, and Saturn) as they appeared to move among the stars. Ptol-

Ptolemy's system. Note how Mercury and Venus revolve about centers that lie on a line from the Earth to the Sun.

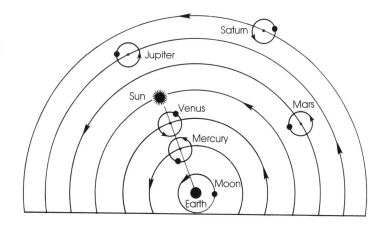

emy devised a complicated and ingenious system for describing these planetary motions. He placed the Sun, Moon, and the "wandering planets" on circles that rotated every day about the Earth, at the center. Each body then revolved about a point on its circle, according to rules that only a mathematical mind would care to follow. Ptolemy's geometric system was an intellectual and practical triumph, even though its predictions were crude by our standards, but no more so than were the observations that were made by the naked eye and timed with sundials and water clocks. Comets, of course, could not be included in his system as their appearances could not be predicted. By the thirteenth century, Ptolemy's beautiful theory had become rather rusty and creaky after some 1,400 years of continuous use, even with repairs. So many serious errors had accumulated that tables of corrections were needed along with tricky patching of the theory itself. In 1252, King Alfonso X of Spain had the tables recomputed; he is said to have told the astronomers that "if he had been present at the creation he would have given some good advice."

Comet observations still remained erratic and imperfect. A comet's appearance changes so much with time that even the most detailed descriptions are almost valueless for identification. Only the position of the comet among the stars at a given time and its apparent motion can identify it. Better and more systematic observations of comets were necessary to disclose their nature.

In summary, then, the ancients and the thirteenth-century thinkers and religious leaders had, for a variety of moral and philosophical reasons, put the heavenly bodies on invisible, crystalline spheres that made mechanical sense, at least to them. This explanation left no place for the erratic comets to move about, except inside the smallest sphere, that of the Moon. Thus, it was concluded that comets must "belong" near the Earth, in the high atmosphere to which Aristotle had delegated them so long ago.

An intellectual explosion was about to blow these crystalline spheres to smithereens—along with a multitude of old ideas and customs—in the Renaissance and its aftermath.

Chapter 3 Breaking the Crystalline Spheres

In about 1238, an unknown writer noted that bright comets tend to be seen near the Sun—that is, in or close to morning or evening twilight. Then, around 1531, the Europeans Jerome Frocastor and Peter Apian observed that comet tails point generally away from the Sun, even ahead of the comet as the comet recedes. The idea was new in Europe, even though the Chinese had remarked on this peculiarity of comets some 700 years earlier. These physical clues to comet behavior later led to the realization that there are vital connections between the Sun and comets. Otherwise, our comet detectives in the thirteenth to fifteenth centuries added little to our story, except for careful descriptions of several bright comets.

On February 19, 1473, however, an important event occurred for the scientific and intellectual world: Nicholas Copernicus was born in Torun, Poland, northwest of Warsaw. Copernicus was the first to propose, on valid scientific grounds but without proofs, that the Sun is the

Peter Apian's woodcut depicting the comet of 1532, showing that comet tails always point away from the sun.

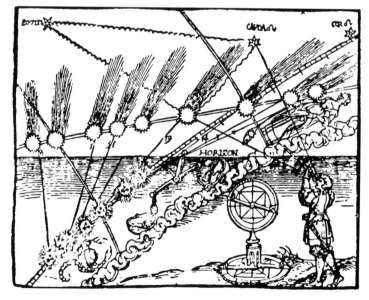

The tails of comets generally stream away from the direction of the sun.

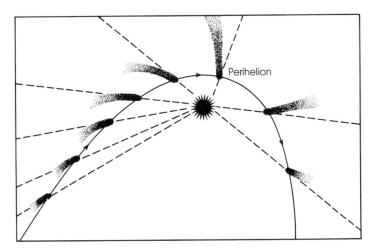

Perihelion

The Copernican system finally put the Sun at the center, with only the Moon revolving around the Earth.

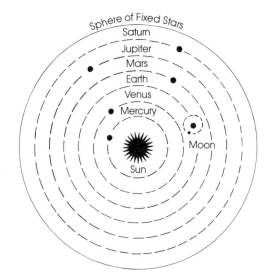

Sphere of Fixed Stars
Saturn
Jupiter
Mars
Earth
Venus
Mercury
Moon
Sun

real center of the Solar System, that the Earth and planets are in motion about it, and that the Earth actually rotates daily on its axis. Because the Church considered such ideas heretical, Copernicus simply showed that, mathematically or geometrically, such a theory gave a better representation of planetary motions than did the cumbersome Ptolemaic theory. Copernicus kept his writings on the subject to himself until nearly the end of his life, but his students spread word of his opinions among scientific circles. Martin Luther, himself one of the greatest reformers of all time, described Copernicus as a fool for holding such opinions, because they were obviously "contrary" to the Bible. Luther considered comets to be harlot stars and works of the devil.

Copernicus's great work, *De Revolutionibus Orbium Coelestium (On the Revolutions of the Celestial Spheres)*, was finally printed at Nuremburg and a copy handed to him on the day of his death in 1543. Several of his friends made the publication possible and introduced some changes in it to appease the Church. The book presented this revolutionary theory as purely a mathematical model having no physical meaning with regard to the actual motions of the planets. Nonetheless, successive generations of students and scholars privately wondered whether Copernicus was really right and also tried to think of ways to prove or disprove his idea. The proof was no easy task. In fact, the stars are so distant that the Earth's motion about the Sun each year displaces them less than a thousandth of the Moon's apparent diameter, an amount that is quite unmeasurable without a powerful telescope. The seed that Copernicus had planted, however, was slowly germinating, not to break out in full vigor for nearly a century.

We should note that, by the time of Copernicus, educated Europeans knew that the Earth is shaped like a ball. The methods of finding distances by triangulation were known even before the time of Eratosthenes of Greece, who, in the third century, B.C., proved to his satisfaction that the Earth was round. He measured the altitude of the Sun at noon on June 21 at Alexandria, when the Sun at Syene in Upper Egypt was directly overhead. The 7.2-degree difference led him to a good value of the Earth's diameter when the distance between Alexandria and Syene had been measured by the travel times of camel caravans.

Myth holds that Christopher Columbus was almost a lone wolf, crying out among flat-Earth skeptics that the world is round. Columbus was actually a super salesman, selling an inferior product, namely, the idea that he could

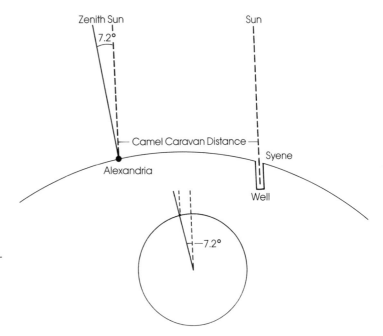

Geometry of Eratosthenes' method for measuring the diameter of the Earth, which should be 16 times the distance from Alexandria to Syene.

sail to the Orient only 2,000 miles west of the Canary Islands. He used a generally unaccepted value of the Earth's diameter that was much too small. His overoptimistic estimate led to a magnificent discovery, but we should not underestimate the knowledge of fifteenth-century Europeans.

The first actual breakthrough in our comet saga was made by the brilliant and flamboyant astronomer, Tycho Brahe. The son of a Danish nobleman, he first studied rhetoric, philosophy, and astrology, followed by mathematics, astronomy, law, alchemy, and medicine—he was a well-rounded student indeed! He is famed for wearing a gold and silver plate on his nose, to cover the part that had been cut off in a duel; this act was illustrative of his irritability coupled with pride and haughtiness. The composition of the nose-piece came into question, however, after his body was exhumed in 1901. The nose-piece had become only a greenish stain. Nevertheless, I like to believe that such a proud man really wore a gold nose-piece and that it was stolen and replaced with a cheap substitute during internment. Tycho's first break with scientific tradition came with his careful observations of the brilliant nova, or new star, in 1572. At first it was as bright as Venus, and then it gradually faded out during the following sixteen months. He showed that the new star was fixed among the other stars in the heavens. But stars were supposed to lie on a crystalline sphere that could allow no changes or newcomers! Tycho had successfully challenged the principle of celestial incorruptibility.

Tycho's fame attracted the attention of the King of Denmark, Frederick II, a zealous patron of science and literature, who in 1576 gave Tycho the rights to the little island of Hveen (now Swedish), along with money to build and maintain an observatory. Tycho built and lived in grand style, but, nevertheless, preserved his scientific integrity. By this time, the errors in the correction tables to the Ptolemaic system of planetary motions were becoming unmanageable. A close approach (or *conjunction*) of Jupiter to Saturn in 1563 was missed by several days in the corrected tables. Tycho's real purpose was to put the

Tycho Brahe. (Reprinted from Petro Gassendi, Tychonis Brahei, Equitis Dani, Astronomorum Coryphaei, vita, 1655, *frontispiece; courtesy Owen J. Gingerich.)*

whole prediction business on a new sound foundation by making the best observations ever of the Sun and planets. At heart, he was a staunch Aristotelian who considered Copernicus to be a rather poor observer.

No modern scientist has been more meticulous than was Tycho in designing and building his instruments. The telescope was yet to be invented, so he had to use only his naked eye with pointers. Tycho developed a completely

Eight drawings of individual comets by Johannis Hevelius showing a variety of tails. Note the apparent sunward antitail of the comet of 1590.

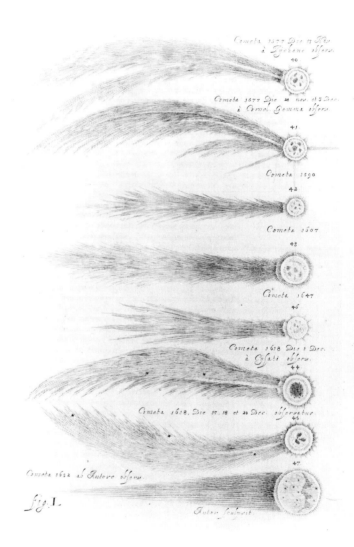

new scientific concept: calibration. That is, he calibrated the errors of his instruments and methods. Because of this innovation and his second innovation of making *systematic* observations of planets, I. Bernard Cohen, Harvard historian of science, credits Tycho with ushering in the era of modern observational astronomy. We know of no individual before Tycho's time who observed a planet every night that it could be seen. Previously observations were carried out only sporadically, to check a new orbit or theory. Tycho's systematic, precise, and long series of observations of Mars made possible Kepler's great breakthrough in planetary motions, as we shall soon see.

In 1577 a brilliant comet set off another frenzy of hysteria in Europe, but Tycho was undismayed by the superstitions. He wanted to find out how distant the comet was. If atmospheric, it would be inside the crystalline sphere of the Moon. If a true heavenly body, it would be beyond the Moon. The Moon's distance was then quite well-known, the Greek astronomers already having placed it correctly at about thirty times the Earth's diameter. Tycho's task was then to compare the direction of the comet on the sky, as he measured it from Hveen, with measures by other astronomers in other parts of Europe at the same time. He concluded that the comet was at least four times the Moon's distance, a vast underestimate of the true distance but quite enough to demolish forever the idea of crystal-

Measuring the distance to the Moon. Observers A and B at positions a and b on the Earth see the Moon displaced among the stars.

line spheres. Other astronomers confirmed the astounding result. Conservative skeptics, however, if they conceded at all, chose to assert that this might be true for the comet of 1577, but not for all comets; they might admit one renegade among the celestial spheres, but not many. The famous French philosopher René Descartes found a way to keep the celestial sphere without violating the comet observations of 1577. He rebuilt the sphere with a myriad of small vortices in a symmetrical pattern, such that the comet could slip through the vortices without cracking the sphere. The literature of the times shows how difficult it is to challenge basic assumptions.

Tycho's total contribution to our story is even greater than this critical proof that comets are heavenly bodies at vast distances from the Earth. As mentioned above, his

Comet of 1577. (Photo by Owen J. Gingerich; courtesy Zürich Zentralbibliothek.)

observations of Mars and the other planets were so precise and systematic that they laid the foundation on which his followers could build a new astronomy.

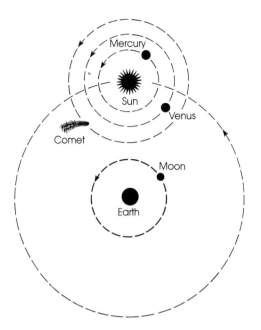

Tycho Brahe's system of the Moon and Sun, with Earth, Mercury, and Venus revolving about the Sun rather than about a line from Earth to the Sun.

The next master mason to add to this intellectual edifice was Johannes Kepler, born in Württemberg, Germany, six years before the comet of 1577. He first studied theology, but at age twenty-three he hesitantly accepted a teaching post in mathematics. There, outside his regular duties, he produced a yearly calendar, or almanac, containing various astrological entries along with predictions not only of the planetary positions, but also of the weather and lucky and unlucky days. Kepler became a popular local prophet, really an astrologer, but unfortunately he was also subjected to religious persecution because he was a Protestant. Finally, this led him to join Tycho Brahe. He then became the heir to Tycho's remarkable observations, particularly those of Mars. Kepler tried all the mathematical theories he could devise to fit Tycho's observations. His confidence in Tycho's accuracy of measurement was the secret of his success. He would not accept a theory that did not predict the direction of Mars among the stars to within about a tenth of the Moon's diameter.

After countless trials using weird numerological and

How to draw an
ellipse.

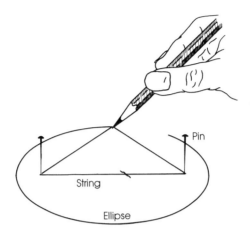

String

Pin

Ellipse

mathematical formulas, Kepler found that the best fit to
the orbit of Mars about the Sun was an *ellipse.* An ellipse
is easy to draw. To do so, stick two pins firmly on a piece
of paper and find a loop of string large enough to extend
beyond the two pins. With a pencil or ballpoint pen, draw
the oval curve that results when the string is kept taut
around the two pins and the point of the pencil or pen.
This is an ellipse with a *focus* at each of the pins. The sum
of the distance from the two foci to any point on the
ellipse is obviously constant; this is how a mathematician
defines an ellipse. In space, Kepler placed the Sun at one
focus of the ellipse and let Mars move along the ellipse
itself.

   Satisfied that planets move in elliptical orbits about the
Sun, Kepler went on to discover two other laws for their
motions. The first, known as the *Law of Areas,* states that
the line from a planet to the Sun sweeps out equal areas
in equal times; the planets move faster when nearer to the
Sun (fastest at *perihelion* when closest, and slowest at
aphelion when farthest away). The second, the *Harmonic
Law,* states that the squares of the periods of the planets
are proportional to the cubes of the long axes of their
ellipses. Kepler had struck a scientific bonanza with his
laws of motion for the planets, even though he never lived
to realize how much he had changed mankind's view of
the universe.

   It is important to our story to note that mathematicians

consider ellipses to be only one example of a class of plane curves called *conic sections*. These curves can be derived by slicing a right-circular cone (a mathematically perfect ice cream cone) with a plane cut. A diagonal slice gives an ellipse. A right-angle slice gives a circle, which is the limit of the hand-drawn ellipse with the two pins stuck in the same place. A slice parallel to the edge of the cone gives a parabola that is open ended, and a slice at a greater angle gives a hyperbola, which is also open ended. The importance of these curves will soon become evident.

Kepler wrote a treatise on comets in which he included the 1607 appearance of Halley's comet. Although Kepler agreed with Tycho that comets are celestial bodies, he held that they move in straight lines to infinity. Some sus-

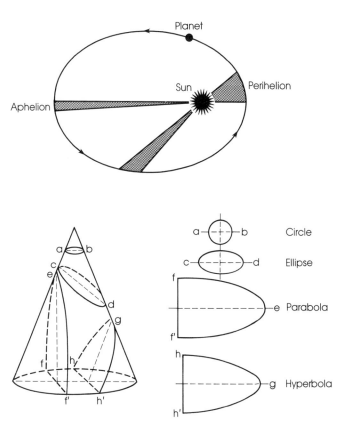

Equal times of planetary motion. Kepler found that planets move fastest when closest to the Sun at perihelion *and slowest when farthest away at* aphelion, *their lines to the Sun sweeping equal areas in equal intervals of time.*

Conic sections: showing how the circle, ellipse, parabola, and hyperbola can be cut from a cone.

pect that Kepler believed comets were not worthy of serious study because of their transient nature; if that was the case, he may have still been under the influence of Plato's philosophy. Evidently he did believe that comets are numerous: "There are as many arguments to prove the annual motion of the Earth round the sun as there are comets in the heavens." He also noted that comet tails point away from the Sun, and then made the extraordinary suggestion that the rays of the Sun penetrate the body of the comet and carry away some portion of its substance. We shall return later to this prophetic physical concept.

During Kepler's lifetime, the heretical belief of Copernicus—that the Earth might be turning and moving about the Sun—was slowly being accepted by European intellectuals. Tycho, even though his observations had cracked the crystalline celestial spheres, never doubted that the Earth was the stationary center of the universe. Copernicus's practical theory was really makeshift, compared with Kepler's. Copernicus used only circles in describing planetary motions, so he had to displace the Sun from the center. As we have seen, Kepler placed the Sun at the focus of a planet's elliptic orbit. It is Kepler's theory that we often ascribe to Copernicus.

Another genius was needed to make Copernicus's idea plausible to philosopher and layman alike. Galileo, born in Pisa, Italy, in 1564, became perhaps the first true physicist in the modern sense. His famous experiments with falling bodies introduced another heretical notion—that material bodies could move freely in space if there was no friction to impede their motion. He actually timed the speeds and accelerations of balls rolling down inclined runways to discover the vital laws describing such motion, in prelude to Sir Isaac Newton's great synthesis of the ideas contributed by Copernicus and Kepler. Whether or not Galileo actually dropped balls of different kinds and sizes from the leaning tower of Pisa, he firmly believed and derived evidence to prove that a feather and an iron ball in space would fall together, following the same laws of motion. In 1610, Galileo heard about someone making a telescope with spectacle lenses. He immediately proceeded to make one himself so that he could look

Galileo Galilei.

at the Moon, stars, and planets. His first telescopic heresy was to find that the Moon was not a perfect sphere, as all heavenly bodies were supposed to be. It appeared to have mountains and craters. Then he saw that Venus was also imperfect; it showed phases like the Moon, proving, to anyone who would look, that Venus must be a ball lighted by sunlight as it moves around the Sun. And, most disturbing of all, he saw that Jupiter was not a lone planet: it had four moons revolving around it.

Thus Galileo, with the help of an important technological breakthrough, destroyed the old Greek idea that the Moon and planets are perfect heavenly bodies, just as Tycho had destroyed the idealized crystalline spheres.

Among these great innovators of the sixteenth century who transformed our view of the universe, Galileo was the one who suffered most because of his beliefs. In his late years, when he was old and sick, Galileo was brought to trial at the hands of the Inquisition. In Rome he was forced to "abjure, curse and detest" the errors of Copernicanism. He spent the last nine years of his life—he died in 1642—in ill health and essentially under house arrest at

The comet of 1680 is depicted here as it appeared in the sky with respect to the constellations, from a contemporary drawing. According to legend, this comet appeared after a hen in Rome laid an egg that was unusually large and marked with the image of a comet. Note the drawing of the egg in the lower right corner. (Photograph by Owen J. Gingerich; courtesy Adler Planetarium.)

Arcetri, Italy. Not until 1983 did the Roman Catholic Church finally "forgive" him.

Galileo, as we have seen, added a critical, realistic factor to what had otherwise been a purely philosophical argument. He, along with Copernicus, Kepler, and others, had accomplished the most difficult task of all in human culture—he had undermined a set of basic assumptions.

Galileo's keen observational skill led him to devise an invaluable instrument for astronomy, the pendulum clock. In 1581, when Galileo was only seventeen, he measured the periods of the swinging chandeliers in a cathedral (counting his pulse??). He found that the period was the same, independent of the arc through which a chandelier might swing in the breeze. He is said to have subsequently designed a pendulum clock, which, however, was never built. The Dutch physicist Christian Huygens constructed the first such clock. It gave the astronomer a method of measuring time with an accuracy commensurate with the increased positional accuracy provided by the telescope.

As for comets, Galileo was drawn into a controversy over three comets observed in 1618. In 1623, he wrote this technically conformist comment:

> Since the motion attributed to the Earth, which I, as a pious and Catholic person, consider most false, and not to exist, accommodates itself so well to explain many and such different phenomena, I shall not feel sure . . . that, false as it is, it may not just as deludingly correspond with the phenomena of comets.

We need not dwell on Galileo's unhappy later years. Suffice it to say that the scientific method had caught on: If one can measure something, devise a mathematical law to fit the measures, and then correctly predict new measures with the law, such a law will outweigh the opinions or dictums of the ancients, the philosophers, the Church, or other authorities. At the same time, however, science is only concerned with things that can be repeatedly measured. The individual is free to speculate or believe whatever he wants about the unmeasurable. Now and then, of course, the distinction between the measurable and the unmeasurable becomes a bit vague, and thus cannot always be clearly defined either by scientists or by laymen.

Chapter 4 **Halley and His Comet**

In the seventeenth century, the telescope suddenly transformed astronomy. Not only had the Moon and planets grown into huge, real bodies, but they could be tracked a hundred times more accurately than with the naked eye and pointers. No rough theory could be accepted now that predictions could be checked on the sky to a thousandth, not just a tenth, of the Moon's diameter. Serious students of nature could dismiss superstitions about the planets and the heavens as irrelevant. Astronomy and astrology began to part company.

Edmond Halley, born in 1656 in this transition period, played a starring role. At the age of twenty-one, he had already made a reputation in astronomy. Among his other accomplishments, he had measured the positions of stars in the southern sky from St. Helena and published the first star catalogue to be based on observations with a telescope. As a result of this work and his other astronomical studies, he was named a Fellow of the prestigious Royal Society (England) while still a young man. On a continental tour, he observed the bright comet of 1680 at the Paris Observatory. At that time, he believed that comets move in straight lines, as had Kepler and a number of other astronomers. Halley's first research on comets was to collect all the observations from the great observer Giovanni Cassini, who was then the director of the Paris Observatory. Halley's attempt to make the observations agree with motion on a straight line was a colossal failure.

In 1682 Halley married the daughter of an officer of the Exchequer. The rumor that he discovered a comet on his honeymoon is pure fiction, but early in his marriage he did observe the comet of 1682 from his private observatory near London. This comet, even brighter than the one of 1680, again excited his interest in comets and made him even more impatient to understand how comets really move. Halley had independently concluded that the force exerted by the Sun on the planets to keep them in orbit was an inverse-square-distance law, but he could not prove it mathematically. This law states that if the distance (r) of the planet from the Sun is doubled, the Sun's attractive force is reduced by a factor of four, or: force varies as $1/r^2$. Neither could Christopher Wren prove the law. He was the astronomer who later became famous as an architect; he is particularly noted for redesigning St. Paul's Cathedral in London after the great fire of 1666. Even the great physicist Robert Hooke was baffled by the mathematics. Wren offered a small prize to anyone who could solve the problem. This incentive brought no results.

Knowing of Newton's genius, Halley went to Cambridge in 1684 to ask Newton about planetary motions. Newton recalled that he had solved the problem some seventeen years before, but had not bothered to publish his solution. Halley was overjoyed at this news and urgently implored the shy, retiring Newton to publish it immediately. The story of the actual publication is a fascinating one of science, diplomacy, and finances. Some of the highlights have to do with Halley's great efforts, not only in pressing Newton to assemble his notes for publication, but also in assisting him with the preparations. When, two years later, the manuscript was ready, the Royal Society of London had no funds left to cover the cost of publication because it had spent so much on *A History of Fishes*. Since Halley's father, who died in 1684 (possibly by suicide), had previously lost his moderate fortune, Halley himself was in an almost precarious financial state. Newton, who may or may not have been able to afford the expense, was in no humor to provide the funds. Finally, Halley managed to assemble enough money to finance the venture. To do so, he had to resign his Fellowship in the Royal Society and obtain a salaried position as clerk to the secretaries of the society. Newton's immortal *Principia* was thus published in 1687.

In the *Principia*, at last, the whole system of planetary motions was developed on the basis of the law of gravitation, the inverse-square law. Newton reasoned that "comets are a sort of planet" and therefore they, too, should move according to the same law. The comet of 1680 had

Sir Isaac Newton.

stirred up a great controversy because it had made two appearances. Having been discovered in the morning sky in November, it disappeared toward the Sun and then was rediscovered in the evening sky during late December. Were there one or two comets involved? At first, Newton doubted that the two were really one, but finally, in his *Principia,* he agreed that they were. In the meantime, pastor Georg Dörffel had made the connection between the two appearances of this comet from his observations in Germany and calculated a parabolic orbit for it. But Newton and Halley may not have been aware of his calculations.

How times have changed! Today, comet discoveries are made known around the world within hours. News of their orbits quickly follows. Centuries ago, many scientists were content to hoard their conclusions, not printing them for many years, perhaps until they had completed a massive tome. Some simply left their manuscripts for posterity. These habits sometimes led to acrimonious disputes as to the original author of an idea or a scientific result. Newton's delay in publishing his work caused just such a controversy with Hooke, who in 1679 had written Newton about the idea of solar attraction and the possibility of an inverse-square law of force.

Many scientists had probably thought about an inverse-square law, but it required Newton's genius as a mathematician to show exactly how an inverse-square force was related to Kepler's law of motion. Furthermore, Newton generalized the inverse-square force of the Sun on the planets into the force of universal gravitation. Whatever the truth of the story about Newton and the apple, he came to realize that gravity on Earth and the motion of the Moon about the Earth represented the same law of force as that for the Sun and planets, and for Jupiter and its moons. This led him to apply the law to the motions of comets and also to the effect that the planets might have on each other and on the comets.

In fact, Newton used his mathematics to make these vital deductions from Kepler's laws of planetary motion. Kepler's laws showed that a force is directed toward the Sun and that it varies as the inverse square of the distance—which is Newton's universal law of gravitation. Galileo's revolutionary idea—that bodies can move freely

Edmund Halley, age 56. Painted by Thomas Murray when Halley was at Oxford. (Courtesy Bodleian Librarian.)

in space—was, of course, the basis for Newton's thinking.

Don't think that Newton's great discovery made a universal hit in the seventeenth century. Action at a distance? How could the Sun reach out to control the motions of the planets? Frankly, I don't understand it myself. Even Einstein's improved concept in his theory of relativity, developed during my lifetime, does not help much. How can matter actually distort space itself? I accept these theories, not because I like or dislike them, but simply because they fit a host of careful measures with incredible accuracy.

Halley had a great head start in understanding comet orbits, but his inherent talent and amazingly wide range of interests kept him away from the problem for some eighteen years. During that time, he had edited the *Philosophical Transactions,* acted as secretary of the Royal Society, and become Deputy Comptroller of the Royal Mint. Earlier he had published a map of the trade winds, which some have thought to be the earliest meteorological chart. His careful studies of the deviation of the compass from true North, so important for worldwide navigation, led to his command of the *Paramour Pink,* a ship devoted entirely to scientific voyages, considered to be the first of the kind. He survived near mutiny, fog, and icebergs at 50 degrees south latitude (Cape Horn is about 56 degrees south latitude), temporary arrest, near shipwreck, and finally even gunfire, when he was mistaken for a pirate near Newfoundland. From the observations made on the voyages, he published *A New and Correct Chart Showing the Variations of the Compass in the West-*

Halley's Map of the Trade Winds, 1686.

ern and Southern Oceans, one of the most important publications in the history of cartography.

In 1704, Halley was elected Savilian Professor of Geometry at Oxford University. Even though Halley attained
this academic post, he was by no means the ideal prototype of academia. His sea experience, or perhaps his
interest in the sea, gave him mastery of "quarterdeck"
language and a fondness for brandy that may have
clouded his later years, from 1720 on, after he had left the
honored post of the second Astronomer Royal. The first
occupant of that post, John Flamsteed, gave Halley little
support for the Oxford appointment, testifying that Halley "now talks, swears and drinks brandy like a sea cap-
tain." This was in 1704.

Furthermore, Halley had unorthodox religious views
that probably would have led to his undoing had he lived
in Europe during the Inquisition. His beliefs probably
prevented him from being appointed to an Oxford Chair
in 1691, a post that he wanted and one that could have
given him an opportunity, some thirteen years sooner, to
pursue his cherished goal of studying comets. As Geome-
trician at Oxford, however, he began to do so in earnest.

Newton had shown that an open parabola can be used
as a close approximation to a long elliptic orbit for a
comet, and indeed, he devised a method of calculating one
from observations. Halley, with incredible patience, compiled the recorded information on twenty-four comets
that had appeared between 1337 and 1698, and from these
data he calculated their orbits, that is, their paths in
space about the Sun. The process is complicated, because
the observer is on a moving platform, the Earth, and the
comet follows an unknown path in an unknown plane that
passes through the Sun. Imagine doing such laborious
calculations without the aid of computing machines! At
least Halley could use logarithms, invented by John
Napier near the beginning of the seventeenth century. The
problem is to find the six unknown quantities, or *orbital
elements,* that will fix the comet's position in space at any
time. The assumption that the orbit is a parabola, not an
ellipse, greatly simplifies the calculations by reducing the
number of unknowns from six to five. For many comets

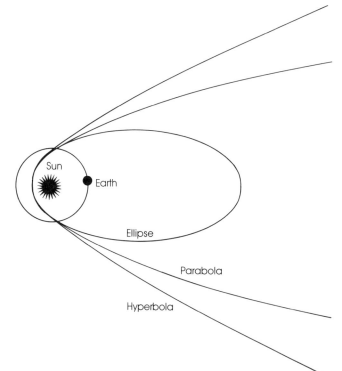

observed for a short time, particularly those in long elliptical orbits, the difference between an elliptical and parabolic orbit can scarcely be detected. This is still true
today with far more accurate measures than were possible in Halley's time.

Two angles on the sky are needed to define the plane of
the orbit, measured with respect to the Earth's orbit
about the Sun. Then the size and the shape or *eccentricity*
(if not a parabola) of the orbit must be found; this step
produces two more elements. Finally, the date of the closest approach to the Sun, or perihelion, must be calculated, and then the direction of perihelion in the orbital
plane. This completes the six elements. Fortunately for
today's comet watchers, the modern electronic computer
has at last taken the drudgery out of orbit calculations.

As Halley compared the twenty-four parabolic orbits of
his twenty-four comets, he was delighted to find three
comets that moved almost exactly in the same plane and
came to perihelion in nearly the same direction from the

Sun at about 0.58 times the Sun's distance from the Earth. These three comets of 1531, 1607, and 1682 also moved backward when compared to the Earth's motion, their orbital plane being tilted only about 18 degrees to that of the Earth's orbit. Could they be the same comet returning with a period of some seventy-six years? Or were there three different comets all moving along the same path? The probability that these were coincidences seemed almost nil.

When Halley looked over the older records of bright comets, he found several that might have been other apparitions. The dreaded comet of 1456, which had worried the Pope, came 75 years earlier than that of 1531. The famous comet of 1066 had appeared 390 years still earlier (5 × 78 years). Halley was bothered for some time by the fact that the periods did not come out exactly the same. After much study, he reasoned that great planets such as Jupiter and Saturn might attract a comet enough to change the comet's period from one apparition to another. He then predicted that the comet would return in 1758, by which time he would have been 102. Although he could not live to see his prediction come true—he died at 85— undoubtedly he never questioned the correctness of his theory, which, indeed, was to be proved correct.

This "Horatio Alger" account of Halley's prediction glosses over the struggles and the mistakes that littered his way to fame in the world of comets. Newton believed

Halley's comet in August 1531, as depicted by Peter Apian.

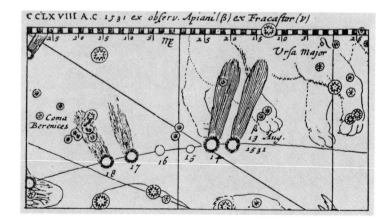

Orbit of Halley's comet projected on the plane of the Earth's orbit about the Sun. Note that the comet goes the "wrong way," or retrograde.

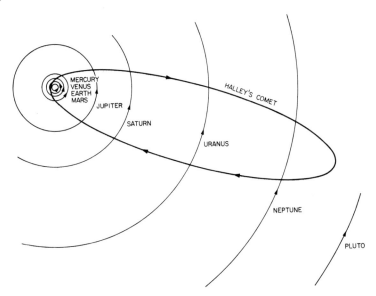

that comets moved in very long ellipses with very long periods, deviating only slightly from parabolas. In Halley's impetuous urge to find examples, he suggested that the comet of 1680 was the return of one seen in 1106, in 531, and in 44 B.C., and that its period was about 575 years. Had his announcement that it would return in the year 2255 been his only prediction, Halley's name might never have been mentioned in the history of comets.

The path of science is a series of ventures into an almost impenetrable maze, with countless dead ends, retreats, false starts, and disappointments. Fortunately, scientists' mistakes are usually interred in archives or else burned. Those of us who make more mistakes than Halley did and discover far less can only admire this extraordinary man with remarkably broad, deep, and almost uncanny scientific vision. His discoveries in Earth magnetism and cartography, his invention of deep-sea diving equipment, his discovery of stellar motions on the sky, and his other contributions to science and to the culture of his time, ensure him a revered niche in the history of science. It is only fitting that his name should be attached to so famous a comet.

Chapter 5

The Returns of Halley's Comet

Man's universe virtually exploded near the beginning of the eighteenth century. That explosion is critical to our understanding of comets. Galileo's idea that bodies can move freely in open space and not slow down without some force to stop them, combined with Kepler's theory of their actual orbits about the Sun, made Copernicus's idea plausible: The Earth might really be a spinning ball, and it might really be plowing through space in an orbit around the Sun. Newton's universal law of gravity added a superb unifying factor, a simple formula that could combine all the observations of heavenly bodies into a "simple" picture of the Solar System, as we know it today. The telescope added the finishing touch: With the accuracy of measurement increasing from a tenth of the Moon's diameter to one thousandth, the theory could be checked to great accuracy and the distances to the Moon and Sun could be measured well enough to define the size of the system. The Earth, fortunately, is big enough to be a baseline for measuring distances to objects in the Solar System, although its diameter, as seen from the Sun, is only 17.6 seconds of arc, which is equal to 1/100th of the Moon's apparent diameter.

The distances turned out to be colossal, almost incredible: The Sun was 150,000,000 kilometers (93,000,000 miles) away! Jupiter was five times that distance from the Sun, and Saturn twice Jupiter's distance. Now there was space to spare for the comets to move about the Sun, any way and anywhere they pleased, with no crystalline spheres to bar their way. Note that an accurate scale of distances is not critical to the predictions of planetary positions, because relative distances are adequate for most problems. The unit of distance is the mean solar distance from Earth (150,000,000 kilometers), which is known as the *astronomical unit* (AU). Only in the recent Space Age have this unit and other planetary measures been known to an accuracy of nearly one part in a million, typical of the accuracy of planetary direction measures.

Galileo's discovery of Jupiter's moons made possible another exciting measure, the actual mass (or weight) of Jupiter itself, which is more than 300 times the Earth's mass and nearly a thousandth that of the Sun's. These numbers must have chilled the marrow of the eighteenth-century conservatives, many of whom still believed that the Earth was the center of the Universe. When the great French astronomer (and Halley's friend) Cassini discovered the moons of Saturn during 1671–84, that planet was found to outweigh the Earth by nearly 100 times.

Now that the masses of these giant planets were known, it became possible to ascertain the effects of their attraction on the motions of each other, of the small terrestrial or earthy planets (Mercury, Venus, Earth, and Mars), and of the comets. Halley's intuition had been right. The orbits and periods of comets are changed by the planets. Practically speaking, however, the theory and calculations are awesome. It is reported that three French astronomers—J. J. de La Lande, A. C. Clairaut, and Madame N.-R. Lepaute—computed incessantly day and night for several months to predict accurately the return of Halley's comet in 1758. Indeed, because of this intensive effort, La Lande contracted an illness that affected

Halley's comet in 1759. Painted by Samuel Scott. (Courtesy Sir Basel Lindsay-Fynn.)

him for the rest of his life. When Clairaut presented the results of these calculations, he noted that Jupiter and Saturn had seriously disturbed the motion of Halley's comet. The comet would be more than 500 days late because of Jupiter's attraction and another 100 days late because of Saturn's. Thus Halley's comet apparently would not come nearest to the Sun until the middle of April in 1759, instead of in 1758 (still with an uncertainty of about a month). The prediction itself came a bit late, as Clairaut did not complete his calculations until November 1758. The search for the comet had already become an international sport, at least in Europe and England.

The professionals, to their chagrin, were beaten out by an amateur astronomer named Johann Georg Palitzsch, a small farmer who lived near Dresden. With the 8-foot-long telescope that he had made himself, Palitzsch discovered the comet on Christmas Day in 1758, fulfilling Halley's prediction. Clairaut had missed the date of perihelion by only 32 out of some 28,000 days; this prediction was a triumph for Newton's theory, and proof that comets are true rovers of the Solar System. As a tribute to Halley, the comet officially carries his name. We are delighted to accede to his own modest wish: "If . . . it [the comet] should return again about the year 1758, candid posterity will not refuse to acknowledge that this was first discovered by an Englishman!"

The next return of Halley's comet in 1835 was widely heralded. The French astronomer P. G. de Pontécoulant won the prediction contest, the comet being late by only three days. Sir William Herschel's discovery of Uranus in 1781 had helped reduce the errors of the 1835 prediction, because the disturbing effect of Uranus on the comet could now be allowed for in the calculations. The comet was magnificent when seen with the sizable telescopes available in 1835, such as the great lens, 15 inches in diameter, of the Pulkovo Observatory near St. Petersburg (now Leningrad). Sir John Herschel, son of Sir William, observed it from Capetown, South Africa; the fine details of rays and jets seen in his drawings near the nucleus are only now being properly interpreted. The astronomer-mathematician Friedrich Wilhelm Bessel was so intrigued by the fountainlike motions he observed near

Comet Halley in 1835, October 13 to October 25, as drawn by Friedrich Wilhelm Bessel.

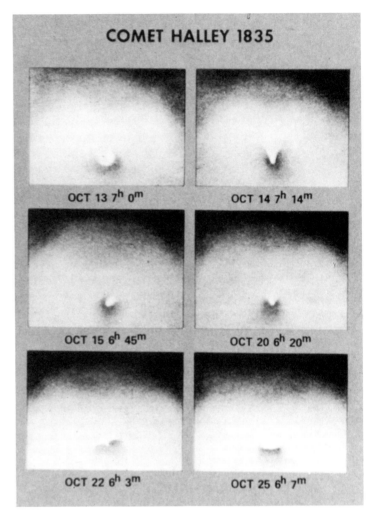

COMET HALLEY 1835

OCT 13 7^h 0^m

OCT 14 7^h 14^m

OCT 15 6^h 45^m

OCT 20 6^h 20^m

OCT 22 6^h 3^m

OCT 25 6^h 7^m

the head of Halley's comet in 1835 that he developed an important theory about comets, one that we shall discuss in detail in chapter 11.

By 1910, the prediction of Halley's comet was expected to be quite exact, to about one day or less in perihelion passage. The last sizable planet, Neptune, had been predicted and then discovered. No massive, hidden bodies were left to disturb the motion of the comet and upset the prediction. The British celestial mechanicians, P. H. Cowell and A. C. D. Crommelin, calculated all the known returns of the comet by the laborious process of adding up the attractions of all the planets, step by step. Their prediction came out no better than Pontécoulant's for 1835, perhaps because his accuracy may well have been somewhat fortuitous.

Cowell and Crommelin wrote, "It now appears from observations that the predicted time, 16.61 April [1910], is 3.03 days too early. At least two days of this error must be attributed to causes other than errors of calculation or errors in the adopted positions and masses of the planets." Their mild disappointment at their lack of hoped-for accuracy was ameliorated in part by a reward of 1,000 marks for the best prediction in what had been a close competition. For our story, this tardiness on the part of Halley's comet is one of our most valuable clues! The comet was actually four days late by present-day precise calculations. Why? The reason finally became clear four decades later, as we shall see in chapter 15.

The return of Halley's comet in 1910 was completely satisfactory to everyone—except possibly to Cowell and Crommelin and certainly to those who expected the comet to produce doom and destruction. Its entrance was upstaged by an unexpected comet that was so bright that no discoverer's name has been attached to it. This comet, known as 1910 I, came rather close to the Sun on January 17, nearly three months before Halley's. It became bright in the evening sky, whereas Halley's comet, when most brilliant in the northern hemisphere, was best seen in the early morning. The two comets are sometimes confused in memory by those who saw one or the other. The tail of Halley's comet stretched out in the sky just before daylight, almost like a searchlight. Photography, well developed by 1910, enabled many observatories to make a beautiful record of the comet, now extremely valuable for understanding its nature.

The coming of Halley's comet in 1910 had received so much advance publicity that many people were truly terrified at the thought of its dire effects. Their fears were amplified by accounts in the press, because certain noxious gases had been discovered in comets and because the Earth was expected to pass through, or at least graze, the edge of the tail.

The unexpected comet 1910 I that upstaged the reappearance of Halley's comet that year. (Courtesy Lowell Observatory.)

Drawings of head of Halley's comet on May 5, 1910, by R. I. A. Innes (left) and W. M. Worsell (right) at Transvaal Observatory, South Africa.

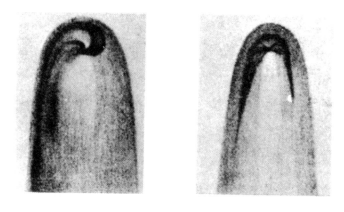

On May 18, 1910, the *New York Times* reported that before sailing on the steamship *Kaiser Wilhelm* to Europe, the actress Olga Nethersole

said that she was not alarmed for herself about what the comet might do to the ship on the high seas, but on account of her pet hairless Mexican dog Chiquita, she was nervous. "You see, if the tail of the comet touched her, my poor pet dog has no hair to protect herself."

On the same day we read in the *Pittsburgh Post*,

Tearfully saying "Goodby" to her friends, a young woman bookkeeper in the Wabash building yesterday went to her employer and tendered her resignation, saying she was going home to her parents in New York, "If I am going to die, I want to be with my mother and father when the time comes," said the misguided one.

In fact, the Earth missed the tail of the comet by a good margin. Thus, two days later the *Chicago Evening American* carried the following story:

Crazed when the comet failed to destroy the earth as he had predicted to his followers, George Ford, giving his address as 3555 Grand Boulevard, attacked policemen and wrecked the Fiftieth Street Police Station patrol wagon when he was being taken to a cell.

Several reports are found of comet viewers who died of heart failure at the sight of Halley's comet. Mark Twain, the great American author, had a characteristically different reaction:

I came in 1835 with Halley's Comet. It's coming again pretty soon, and I expect to go out with it. It'll be the great disappointment of my life if I don't go out with Halley's comet. The Almighty has said, no doubt, "Now here are two indefinable freaks. They came in together. They must go out together." Oh, I'm looking forward to that.

In fact, Samuel Langhorne Clemens was born on November 30, 1835, two weeks after Comet Halley's perihelion passage, and died on April 21, 1910, the day after the comet's closest approach to the Sun!

On the lighter side, the *Chicago Tribune* in May 1910 ran the following story: "At a special meeting of the General Committee for Reception of Halley's Comet, Professor Graham Taylor read a report from his Oxford colleagues." The report outlined the following experiment: "Empty 24 champagne bottles and fill them with Halley Comet cyanogen gas for future scientific tests." This, to my knowledge, was the first proposed astronomical experiment. It failed, of course, because the Earth missed the tail of the comet, not because of any failure of the experimenters to empty the champagne bottles.

Others were less impressed with the apparition of the comet. In the *Chicago Daily Tribune* on May 26, we find:

We wish to remark that our idea of nothing to see is Halley's comet. . . . An expiring Chinese lantern with its candle guttering in the last gasp at a church sociable and strawberry fete after the festival folk had gone home and the janitor had stacked up the chairs would present a magnificent and awe inspiring spectacle compared with this gaseous fraud.

This attitude, I fear, may well represent the reaction of many people to Halley's comet during its 1986 reappearance. At the time when it comes closest to the Sun at perihelion (1986 February 9.7) and is intrinsically almost its brightest, it lies nearly in the direction of the Sun but beyond it, and is practically unobservable. Its more favorable appearances to Earthlings before perihelion in late December 1985 and after perihelion in March and April 1986 are unfavorable, except possibly for observers with binoculars located away from cities and surrounded by very clear skies. Observers in the Southern Hemisphere have a better view of P/Halley in April, however, as it is

A 1910 German post-card cartoon picturing a comet hitting the earth. The earth actually passed through the outer portion of the tail of Halley's comet in 1910. (Courtesy Donald K. Yeomans.)

then easily seen with the naked eye high in the morning sky.

D. C. Jewitt and G. E. Danielson headed an observing team that recovered Halley's comet on October 16, 1982, with the 200-inch Palomar reflector. Two nights later, M. J. S. Belton and H. R. Butcher at the Kitt Peak National Observatory confirmed the recovery with the 4-meter reflector. In so doing, these observers broke two records of cometary recovery: (a) distance from the Sun, 11.05 AU; and (b) faintness, 1/19,000,000th of the faintest limit by the naked eye.

Fortunately for scientific studies, we are assisted in 1986 by the power of the Space Age to overcome some of the limitations imposed by our location on this small planet and its hazy atmosphere (chapter 24).

Because Halley's comet is unique in being the first to be recovered through prediction, because various of its thirty apparitions have gained worldwide attention, and because it has contributed so much to the understanding of the physical nature of comets, a sequential account of these apparitions may be of interest. The following list is arranged in the order of the perihelion passages, which are based primarily on the calculations of Yeomans and Kiang and of Hasegawa.

240 B.C., May 25   Best substantiated as the first recorded observation, made by the Chinese. According to Ho Peng Yoke, University of Maylaya, Singapore, "During the seventh year of Chhin Shih-Huang-Ti, a (hui) comet first appeared at the North and during the fifth month it was seen at the West. [Later] it was again seen at the West." Earlier observations of P/Halley in Chinese records were suggested by Y. C. Chang of the Purple Mountain Observatory, China, in 1978. These perihelion passages for which Chang finds supporting evidence were in 1057 B.C., 615 B.C., and 466 B.C., although the historical records are difficult to interpret.

164 B.C., November 12   Observed at Rome, and probably in Japan, this is the first and only record of Halley's comet in Babylonia. Found by David Pingree of Brown University.

87 B.C., August 6   Observed by the Chinese in autumn in China and also in Rome.

12 B.C., October 10   The Chinese followed it in the constellation Gemini for fifty-six days, from August 26 until it disappeared in the constellation of Scorpius, traveling from the western morning sky into the western evening sky. According to Dion Cassius, a Roman historian, "A comet hung like a sword over Rome before the death of Agrippa (in 12 B.C.)." The Chinese account of the apparition and the constellations through which it passed, along with the descriptions of its physical appearance, contrasts markedly with the Western account.

A.D. 66, January 25   Seen in the east by the Chinese. Quite possibly this is the comet that Joseph described as a sword-shaped sign hanging over Jerusalem, warning of its destruction.

141, March 22   Observed in China for three months with details of its position among the constellations. Also in Europe. Plague was raging in both countries.

218, May 17   The path observed by the Chinese is in excellent agreement with calculations. Dion Cassius describes the comet as "a very fearful star with a tail." It marked the murder of the Roman Emperor Macrinus.

Halley's comet with Venus on May 13, 1910. (Courtesy Lowell Observatory.)

295, April 20   Seen in China for seven weeks.

374, February 16   Seen in China in March and April.

451, June 28   Brilliant in Europe and seen for twelve weeks in China. Associated in legend with the Battle of Chalons, in which Attila, the King of the Huns, was defeated by the Christian armies.

530, September 27   The Chinese found it in the morning sky on August 29 and saw it reappear in the evening of September 4; it finally disappeared on September 27. Merlin, the wizard of King Arthur's Court, is said to have based prophecies on the comet.

607, March 15   The Chinese observed it for at least a hundred days.

684, October 2   Recorded by the Chinese, Japanese, and Europeans. This was the period of the Black Plague.

760, May 20   Well observed in China for more than fifty days and in Europe.

837, February 28   Observed extensively in China and in many parts of the globe, including Japan, from March 22 to May 7. Generally believed to have presaged the death of King Louis le Debonair of France. Closest approach of P/Halley to the Earth: 5 million kilometers on April 15.

1066, March 20   Immortalized in the Bayeux Tapestry. In China it was first seen on April 2, and lasted for sixty-seven days. In Korea, "a star like a moon rose from the N.W. Presently it transformed into a comet" (Ho Peng Yoke). The comet was thus tailless about a month after perihelion. It was also described by Zonares, the Greek historian, who found it to be as large as the full Moon before the tail appeared. Observed in Japan also. Could be seen during full Moon on June 10, and was last seen July 8. An unusual outburst may have occurred.

1145, April 18   Observed throughout the world. First seen in Europe on April 15. The Chinese recorded an enormous tail on May 14, when the comet was in Orion. Seen by the Chinese up to July 9.

1222, September 28   Regarded in Europe as the precursor of the death of the King of France, Philip Augustus. Observed in China, Japan, and Korea from September 7 to October 23.

1301, October 25   Seen worldwide, especially in the north as far as Iceland, with a bright long tail. It inspired the Italian painter Giotto di Bondone to include a comet as the Star of Bethlehem in his famous painting *The Adoration of the Magi*. Interval of observation: September 1 (Europe) to end of October (China).

1378, November 10   Seen in China, Korea, Japan, and Europe. Not especially conspicuous, although quite bright intrinsically. Observed for forty-four days.

1456, June 9   A particularly favorable apparition. Chambers reports that the Chinese described the comet "as having had a tail 60 degrees long and a head which at one time was round and the size of a bull's eye, the tail being like a peacock!" Seen from May 27 until at least July 6 in China. This is the comet that frightened Pope Calixtus and almost everyone else in Europe.

1531, August 26   First observed in Europe at the end of July and seen in China until September 8. Orbit first calculated by Halley and, along with the orbits for the comets of 1607 and 1682, was the basis for his theory of periodicity among comet orbits.

1607, October 27   Discovered by the Chinese on September 21 and first observed by Kepler on September 26 (or 23?). Became a conspicuous object on October 1. Still seen in China on October 26. In the same period, Jamestown was settled in the New World.

1682, September 15   The first apparition of P/Halley to be observed with telescopes. Discovered on August 26 by French astronomers and by the Chinese, all with the naked eye. Last seen on September 22, in Paris, the Chinese having lost sight of it after September 8. The third apparition of the comet was used by Halley in his initial calculations.

"Comet Rag," by Ed. C. Mahony. (Daly Music Publishers, Boston, 1910.)

1759, March 13   The first accurately predicted return of a comet. Discovered on December 25, 1758, by Johann Georg Palitzsch. Observed telescopically until March 18 by Charles Messier.

1835—III, November 16   Beautifully placed for observation. Observed for more than nine months from August 5, 1835, until the middle of May 1836.

1910—II, April 20   Another apparition in which the comet was well placed for observation. Recovered by M.

Recovery image of Halley's comet on October 16, 1892, made with the 200-inch Hale telescope at Palomar Mountain by David Jewitt and G. Edward Danielson.

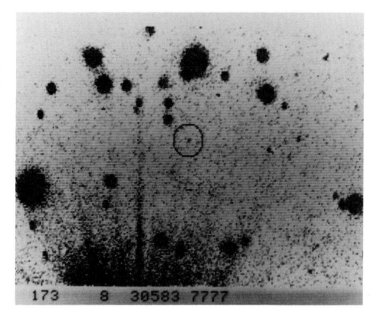

Wolf, Heidelberg. First photographed in Helwan, Egypt, on August 24, 1909, and finally on June 16, 1911. Hence the comet was under observation for more than twenty-one months!

1986, February 9.7 (calculated)   Comet 1982i, recovered by D. C. Jewitt and G. E. Danielson on October 16, 1982, with the 200-inch Palomar reflector using an electronic camera. Magnitude 24.2 visual, or 1/19,000,000 the limiting brightness visible to the naked eye. The solar distance at recovery was 11.05 AU, a record. The error in the position on the sky from the calculations in 1977 by D. K. Yeomans at the Jet Propulsion Laboratory was 8 arcsec, or less than 1/200 the Moon's apparent diameter, amounting to about 0.3 day error in the calculated time of perihelion passage. The actual error, of course, depends upon the observed time of perihelion passage.

2061, July 29 (calculated)   Another unfavorable apparition for observations from the Earth.

Chapter 6   **The Sport of Comet Hunting**

The amateur Palitzsch set no precedent when he recovered Halley's comet in 1758 by means of a telescope. The first telescopic discoverer was apparently Godfrey Kirch of Coburg (now in West Germany), who found the brilliant comet of 1680—the one that had intrigued Halley and was used as an example in Newton's *Principia* (Volume III). As telescopes improved, a few professional and amateur astronomers began to compete for the glory of discovering comets. Perhaps the most avid, if not rabid, of these comet hunters was Charles Messier, who found his first comet in 1759, immediately after failing in his goal of recovering Halley's comet. He went on to discover at least twelve more, for which he receives credit in modern catalogues of well-observed comets. His ardor for comet hunting seems to have matched or even exceeded that for his wife. While he was caring for her during her last illness, he had no time for comet seeking. His rival, Montaigne of Limoges, beat him to the discovery of a comet. It is said

Charles Messier. (Painting in Paris Observatory; photograph courtesy Owen J. Gingerich.)

that, after his wife's funeral, his reply to someone offering condolences was, "Alas . . . Montaigne has robbed me of my thirteenth comet." In consternation at this Freudian misperception, he quickly murmured, "Oh, my poor wife." Whatever the truth of this story, he apparently did fall into a well at night, while looking upward in hopes of seeing a comet. The accident put him in bed for several months.

As a result of Messier's infatuation with comets, his name frequently appears in many current astronomical publications. He repeatedly found that a new "comet" failed to move on the sky, because it was really a hazy, fixed nebula with the appearance of a comet. To remedy this recurring nuisance, he identified, measured, and described 103 nebulae. This catalogue helped Messier ascertain whether or not a hazy patch was really a comet. The great star cluster in Hercules, for example, is M13, standing for no. 13 in his catalogue. He failed to list a number of bright nebulae, however. Much larger, subsequent catalogues, such as Dreyer's *New General Catalogue*, extended the list of fixed, cometlike objects.

The reward in discovering a comet comes primarily from the immediate satisfaction and modest acclaim, plus a form of immortality in comet catalogues. From time to time, the rewards have been more tangible. Between 1831 and 1847, Frederick VI, King of Denmark, gave gold medals to the discoverers of comets, as did the Vienna Academy of Sciences later, until 1900, when the golden era ended. Beginning in the 1880s, H. H. Warner, an American philanthropist, offered a reward of 200 dollars. At that time, Edward E. Barnard, a young and poor photographer, became interested in optics and then in comet hunting. His own account tells the story:

Times were hard in the last of the seventies and the first of the eighties, and money was scarce. It had taken all that I could save to buy my small telescope. After saving and borrowing and raising a mortgage on the lot, we built a little frame cottage where my mother, my wife, and I went to live. These were happy days, though the struggle for life was a hard one. . . . However, the hand of Providence seemed to hover over our heads: for when the first note came due a faint comet was discovered wandering along the outskirts of creation, and the money [from Warner's award] went to meet the payments. The faithful comet, like the

E. E. Barnard with the Lick Observatory's 36-inch refractor in the 1890s (Lick Observatory Photograph).

E. E. Barnard's comet house, Nashville, Tennessee, ca. 1885. This house was built from award money given to Barnard for his discovery of several comets. On porch are Barnard's wife and mother. (Photo courtesy A. Heiser, Dyer Observatory, Vanderbilt University.)

goose that laid the golden egg, conveniently timed its appearance to coincide with the advent of these dreadful notes. . . .

And thus it finally came about that this house was built entirely out of comets. The fact goes to prove further the great error of those scientific men who figure that a comet is but a flimsy affair after all; for here was a strong compact house—albeit a small one—built entirely of them. True, it took several good-sized comets to do it; but it was done nevertheless.

Barnard, in fact, had remarkably acute eyesight and was an indefatigable observer. He worked his way from obscurity to an enviable niche in American astronomy near the turn of this century. He discovered nineteen comets in all, mainly by eye and telescope.

The all-time winner in the hall of fame for comet finding is Jean Louis Pons, who started his career as a doorkeeper at the Marseilles Observatory. Between 1801 and 1828, he discovered, or shared in the discovery of, thirty of the forty-three comets that were observed in that interval and subsequently listed with reliable orbits in modern comet catalogues. He actually claimed thirty-seven comets in all, but several were not observed frequently enough for reliable orbits to be calculated. Pons's record may well stand forever. Today there are many more competitors, including unintentional competitors—professional astronomers who find comets by chance on their photographic plates. Perhaps Barnard was the first to do so in 1892, at the Lick Observatory in California, although a comet may have been photographed from Egypt when it appeared near the solar corona during the total eclipse of the Sun in May 1882.

Among the comet hunters in 1984, two amateurs and one professional are tied for the lead with twelve discoveries each. They are amateurs William A. Bradfield of Adelaide, Australia, and Minoru Honda of Kurashiki, Japan, and professional Antonın Mrkos of the Skalnaté Pleso and Klet' observatories in Czechoslovakia.

The woman with the most comet discoveries is Caroline Herschel, who is credited with eight independent comet discoveries from 1786 to 1797. She used a small telescope built for her by her brother, Sir William. Her contribution to his success is probably greatly understated in astronomical history because brilliant women of her day had

William A. Bradfield of Dernancourt (near Adelaide), Australia, well-known discoverer of comets in the 1970s and 1980s. (Courtesy Dennis di Cicco.)

almost no opportunity to make scientific history. The story is quite different for the late Ludmilla Pajdušáková, who discovered five comets between 1946 and 1954. She achieved the directorship of the Skalnaté Pleso Observatory. Another prolific woman comet observer in recent decades has been Elizabeth Roemer, now at the Lunar and Planetary Laboratory at the University of Arizona. Her innumerable precise photographic observations of comets have been the basis for a great many cometary orbits of importance, and her notes on the physical characteristics of comets have also been invaluable.

Well-known comet discoverer Minoru Honda with some of his observing equipment. (Photograph by Akio Nakamura.)

Antonín Mrkos of Czechoslovakia, discoverer of twelve comets through early 1985.

The late astronomer Ludmilla Pajdušáková, former director of Skalnaté Pleso Observatory in Czechoslovakia, discoverer of several comets.

Elizabeth Roemer.

A prolific comet discoverer among women currently is Carolyn Shoemaker, who has conducted photographic surveys for new comets and close-approaching asteroids with her husband, Eugene Shoemaker. Using the 46-centimeter Schmidt telescope at Palomar Mountain in California, the Shoemakers found six new comets during a fifteen-month span between 1983 and 1984. Two of these are of short period.

Comet discovery usually takes place as follows: The visual observer (male or female) spends many hours over a period of months, or even years, scanning the sky through a wide-field telescope, looking particularly at regions in the evening and morning skies where comets are most likely to appear. He sees many hazy patches, which he checks on his star charts. They turn out to be nebulae. Finally, he finds one that is not a known nebula. He watches it to determine whether it moves among the stars, and consults the comet predictions to see whether it is already known. Clouds, daylight, or the setting of the sky in the west may stop him before he can investigate its motion. If these conditions prevail, the next night is almost certain to be cloudy! By the time he can look again, he must try to rediscover the hazy patch. If the

Typical discovery image. Comet Peltier-Whipple-Sase 1932 V. (Harvard College Observatory.)

hazy patch has moved, he has possibly found a comet. The suspected comet can still turn out to be a "ghost image" or reflection due to the telescope's optics. To discover a comet photographically, the observer follows much the same procedure, except that usually another photograph must be made on a later night (or later the same night) to capture the sequence of movement and thereby confirm the reality of the object and its proper motion. In all cases, the sophisticated comet hunter first makes sure that a nearby bright star, planet, or crescent Moon hasn't produced the "comet" by creating internal reflections in the telescope.

Once satisfied that the "comet" is probably real, and not previously reported, the discoverer communicates his sightings by telegram or telephone to a center designated by the International Astronomical Union (IAU) and administered by a designated (and devoted) orbit expert. In 1985, this astronomer was Brian G. Marsden, who headed the IAU's Central Bureau for Astronomical Telegrams at the Smithsonian Astrophysical Observatory in Cambridge, Massachusetts. Ideally, the message should contain the approximate coordinates on the sky, motion on the sky, date and time of observation, estimated brightness of the object, observer(s), telescope(s) used, and mention of tail (if any). A standardized self-checking code has been established for telegrams.

The bureau then attempts to determine whether the suspected comet is an already known object and also whether it is located near a bright object on the sky. If doubts still remain about the discovery, the bureau sometimes asks for help from a small, select list of cooperating visual and photographic observers, both amateur and professional. Once a discovery is confirmed, the bureau sends out an announcement via telegram to a longer list of interested observatories and individuals. A postcard circular is mailed to a still much longer list of subscribers, including members of the news media.

Frequently, some independent discoveries are made before the news gets around. The comet may then be assigned up to three names of independent discoverers, as well as a year and letter; for example, Whipple-Bernasconi-Kulin, 1942a, designates the first comet dis-

Prolific American comet hunter Leslie Peltier with his home-built backyard observatory in Ohio.

covered (new) or recovered (old) in 1942. A year or two later, after precise orbits have been calculated for the comets observed near that same time, it receives its final Roman numeral designation, in this case, 1942 IV; the Roman numeral indicates that this was the fourth known comet to come to perihelion in 1942. A comet that turns out to have a period less than 200 years is called a *periodic*, or short-period comet and is designated by P/, as in P/Pons-Winnecke. If the discovery turns out to be the return of a previously known comet, as established by its orbital elements, it may still carry its original name. There are also planned "rediscoveries," known as *recoveries*, based on predictions from previous returns. For example, P/Pons-Winnecke was recovered in 1976 by Elizabeth Roemer and C. A. Heller of the Lunar and Planetary Laboratory, Tucson, Arizona. In this case, it carries the

Circular No. 3796

Central Bureau for Astronomical Telegrams
INTERNATIONAL ASTRONOMICAL UNION

Postal Address: Central Bureau for Astronomical Telegrams
Smithsonian Astrophysical Observatory, Cambridge, MA 02138, U.S.A.

TWX 710-320-6842 ASTROGRAM CAM   Telephone 617-864-5758

COMET IRAS-ARAKI-ALCOCK (1983d)

A fast-moving object, originally thought to be asteroidal, was reported by John Davies, Leicester University, England, in the course of the minor-planet survey with the Infrared Astronomy Satellite (IRAS). He reported this discovery to several observers (but not to the Central Bureau) on Apr. 26. According to an unclear recorded message from H. Rickman, Uppsala, the object was confirmed on Apr. 27 by T. Oja, Kvistaberg, Sweden, and said to be a comet, but no positional information was provided. Lacking further data, J. Gibson took plates for the IRAS object at Palomar on May 2.5. Before he could inspect these plates, the Central Bureau received word via G. Keitch that George E. D. Alcock, Peterborough, England, had discovered a possible bright comet on May 3.9, and within an hour confirmation of the discovery was received from G. M. Hurst, Wellingborough, England. Suspecting that the two comets were identical, but still lacking any quantitative data on the IRAS object, the Bureau obtained the IRAS information from Gibson, at which time he noted the comet. On May 4.1 the Bureau telegraphed the Alcock discovery and the indicated motion based on the IRAS observations. Immediately afterward, Y. Kozai, Tokyo Observatory, telexed news of the May 3.6 independent discovery by Genichi Araki, Yuzawa, Niigata, Japan. In response to an urgent request, Davies then provided the IRAS and Swedish observations. Visual observations on May 4 have been made by D. W. E. Green, A. C. Porter and others. Green notes that the comet is extremely large, of diameter 18' (20 x 80 binoculars). Positions follow:

1983 UT		α_{1950}	δ_{1950}	m_1	Observer
Apr.	25.85478	19^h06^m29	$+48°38'.6$		Davies
	25.92649	19 06.15	+48 39.4		"
	27.88773	19 04 56^{s}35	$+49 17 03''.8$		Oja
	27.90643	19 04 55.53	+49 17 30.1		"
	27.93551	19 04 53.94	+49 18 10.5		"
May	2.46120	18 58 27.59	+51 42 09.4		Gibson
	2.49106	18 58 22.79	+51 43 35.3		"
	3.61111	18 56	+52 30	7	Araki
	3.916	18 54.8	+53 02		Alcock
	3.95486	18 54.1	+52 58	6.5	Hurst
	4.23	18 52.9	+53 24	6.0	Green
	4.306	18 52.8	+53 28		Porter

1983 May 4   Brian G. Marsden

temporary designation 1976f, which indicates that it was the sixth comet to be discovered/recovered in 1976. It is also known as comet 1976 XIV, which means it was the fourteenth to pass through perihelion in 1976.

If, as opposed to a planned recovery, a comet is rediscovered after having been lost for a number of revolutions, its second discoverer often receives credit by having his or her name attached. In our example P/Pons-Winnecke, Pons first spotted the comet in 1819, and so it was called comet 1819 III. Its period is 5.6 years. This comet was lost for six returns, but then was independently rediscovered in 1858 (as 1858 II) by F. A. T. Winnecke at the Bonn Observatory, in Germany. When his name was added, the designation became P/Pons-Winnecke. Thus, several such comets carry two or three names, but none more than three.

The case of P/Halley has already been mentioned as one periodic comet named for someone other than the discoverer. There are few such examples, the most notable being P/Encke, P/Crommelin, and P/Lexell, named for the men who made elaborate and valuable calculations of the orbits. Some observatories (Tsuchinshan, China's Purple Mountain Observatory) or artificial satellites (SOLWIND and IRAS) have had their names attached to comets, usually for political reasons. At times, even an observatory director has claimed the name of a comet discovered by an unrecognized assistant. A famous example occurred in 1867, when Jerome Coggia, at the Marseilles Observatory in France, discovered a comet. Eduard Stephan, who was director of the observatory at the time, released the news of the discovery under the name "Comet Stephan." Today the object still bears his name, P/Stephan-Oterma, having been independently rediscovered in 1942 by Liisi Oterma of Finland.

Sky searching for comets seems to require an average of about 150 to 200 hours for each success. I can confirm this estimate from my own experience after looking over the photographic plates taken for sky surveys conducted by the Harvard College Observatories. To discover six comets in twelve years, I scanned some 70,000 glass negatives, 8 by 10 inches in size, with a hand magnifying glass. I spent most of my time identifying emulsion flaws and the pesky nebulae that Messier had found so confusing. My assignment was to inspect the quality of the photography and processing. Comet hunting was a secondary activity, but far more fun. Only one of my six comets, discovered in 1933, turned out to be of short period: P/Whipple. Though faint, it has returned faithfully seven times, every seven to eight years. My tangible rewards for these comet discoveries are six handsome bronze medals, donated by Joseph A. Donohoe and awarded by the Astronomical Society of the Pacific.

A secondary sport to comet hunting was, for about a century, preliminary orbit calculating. Each time a new comet is discovered, its path on the sky is predicted so that other astronomers can observe it. First, however, its orbit must be calculated from a minimum of three well-defined positions on the sky—with respect to the known positions of nearby stars—on at least three successive nights. By the 1850s, many good star positions were available throughout the sky. With the advent of telegraphy, followed by transoceanic cables, the telephone, radio, and finally satellite communication, the transfer of comet information around the world became streamlined. In the old days, the would-be orbit computor waited for the third usable position of the new comet to reach him by coded telegraph. He then began calculating feverishly, hoping to be the first to cable or telegraph the orbit and positional predictions to the "comet center," for worldwide distribution. With logarithms, and later table-model

Joseph A. Donohoe Medal for discovery of Comet P/Whipple in 1933.

mechanical computers, the calculations took six to twenty hours, depending upon one's skill and luck with the solution. The calculations were usually carried out in one sitting, regardless of the starting time. Coffee and sandwiches were vital aids. The record time for computing an orbit without modern computers may be on the order of two hours, but this is far from typical. I have computed more than forty orbits by the old methods, but my shortest time was far greater than four hours. The modern electronic computer has largely killed the sport of preliminary orbit computing by reducing the actual computing time from hours to seconds. Now the difficult task is observing and measuring the comet's position on the sky, all the calculations being trivial once the computer programs have been coded.

But modern technology has not quelled the ardor of comet hunters. Amateurs still make a considerable and much-appreciated contribution to cometary science, not only by finding comets frequently, but also by observing their physical appearance and brightness. Often amateur observers are the first to report an unusual phenomenon, such as a large brightness outburst in a comet, and therefore make highly valuable contributions to the science of comets.

American amateur John E. Bortle (left), well-known comet observer, and Japanese amateur Tsutomu Seki, master comet discoverer and astrometrist.

One amateur who discovered only one comet (1930 IV) deserves special mention for his contribution to the science of comets. He is Max Beyer of Hamburg, West Germany. Between 1921 and 1970, he carefully observed 110 comets using telescopes on the grounds of the Bergedorf Observatory. For each comet, he made many measures of brightness, diameter, and physical appearance, all of which add up to an invaluable library of cometary data. His thousands of meticulous observations have been a boon to me in my own research.

Modern technology and the Space Age have added a new wrinkle to comet discovery. For a century, the methods remained unchanged. The discoverer used his eye with a telescope, or else a photographic plate or film. But in January 1983, the National Aeronautics and Space Administration (NASA) launched a satellite with infrared sensors, the Infrared Astronomical Satellite (IRAS). The designers of IRAS hoped to image all the bright sources of infrared light and heat on the sky and hoped also to discover some of the asteroids or small planets that come near the Earth. To practically everyone's surprise, IRAS turned out to be a comet finder *par excellence*. During its eight months of active life in orbit, IRAS made possible the discovery of at least six new comets and the rediscovery of a number of others already under observation.

Most noteworthy was comet IRAS-Araki-Alcock (1983d). John Davies in England was the first to observe the fast-moving comet in the IRAS data, followed by two independent discoverers, the amateurs Genechi Araki in Japan and George Alcock in England. All three sightings were made in late April and early May. On May 11, 1983, the comet came within 2,900,000 miles of the Earth, or 0.031 AU, closer than any known comet since Lexell's of 1770. This exciting opportunity for a close look at a comet galvanized the astronomical community into action, even though the time for preparations was short (less than a week).

IRAS helped to swell the number of comet discoveries and recoveries in 1983 to a new record of twenty-two. This almost created a small crisis for those assigning preliminary comet designations, because the alphabet runs out at twenty-six. Among the twenty-two comets, twelve

French cartoon of comet discovery. (Courtesy Owen J. Gingerich.)

were new comets and ten were recoveries of periodic comets. A twenty-third letter was assigned in 1983 to a mistake—actually a photographic plate flaw! More comets may be found when the IRAS data bank has been studied in greater detail. Interestingly, IRAS did miss some known comets that it passed over but did not detect. No fewer than five amateur astronomers had their names attached to comets discovered during the eight-month life of IRAS. An intriguing question follows: Will infrared satellites of the future monopolize the sport of comet hunting? Probably not for a long while.

Most comets discovered in the late 1970s and 1980s have been found by professional astronomers who noticed the new, fuzzy objects on their photographic plates; such plates are usually taken during various astronomical surveys for interstellar clouds (including IRAS), asteroids, and numerous other programs. These comets tend to be faint and thus suggest that we are still missing a great many of the periodic comets.

Chapter 7

The Consequences of Comet Hunting

More than two centuries of intensive comet hunting, coupled with information from older records, give us vitally important clues to the habits of comets, how they move, and where they come from. The authoritative catalogue of comet orbits, compiled in 1982 by Brian G. Marsden at the Smithsonian Astrophysical Observatory, contains 1,109 orbits for 710 individual comets observed, a far cry from the 24 so laboriously calculated by Halley. As the measurement accuracy of star and comet positions improved, most comet orbits, surprisingly, turned out to be either very long (close to parabolic) or rather short and oval (closer to a circle).

Indeed, a number of orbits seemed to be slightly open or hyperbolic, as though the comets were visitors to the Solar System from the depths of interstellar space. The orbits of the "wild comets" were tilted every which way compared with those of the orderly planets, which move in the same direction and near the same plane. Quite a

Brian G. Marsden.

Wild new comet, Kohoutek 1973 XII, on January 12, 1974, shortly after perihelion. (Photographed with 48-inch Schmidt Telescope, Palomar Mountain; courtesy the Hale Observatories.)

few comets, on the other hand, revolved about the Sun in short periods of a few years in the same sense as the planets, their orbits generally being tilted a little, up to about 30 degrees, from the plane of the planets. These comets seemed to have been "tamed" somehow by the giant planets Jupiter and Saturn, in that they usually ventured no farther from the Sun than the orbits of these giants.

These orbital clues have now been sorted. The numbers tell the story. Of the above-mentioned 1,109 comet apparitions, only 710 are individual comets, because 121 "tame ones" of short-period account for 520 passages. The other 589 have been seen probably only once each. Somewhat more than half of the 589 have been so poorly observed that parabolas must serve as the best orbits available. There is no way to be certain whether their periods are short or long. This leaves 169 comets that truly move in extremely long-period orbits, of up to a million years or more, and 104 that move in slightly hyperbolic orbits. The evidence is striking. Out of the 121 tame comets, all but 4 revolve around the Sun in the same sense as the planets, that is, in prograde orbits. All but 16 of the 121 have orbital periods less than thirty years, and two-thirds of them go no more than one AU beyond Jupiter. A plot of the orbits of the short-period comets projected on the plane of Jupiter's orbit show a remarkable clustering. The ring of their aphelion curves outlines Jupiter's orbit beautifully. The conclusion has been clear for more than a century! Jupiter's huge attractive mass has somehow collected two-thirds of all the short-period comets into a family. Saturn probably also plays a supporting role in the process. Jupiter and Saturn appear to be much more important in the story of comets than was indicated by their slight disturbances of the motion of Halley's comet. The existence of Jupiter's comet family is one of our important clues to the origin of comets.

Note that Halley's comet is one of the four independent mavericks not dominated by Jupiter; it moves in almost the opposite direction and its aphelion is beyond Neptune. A fifth such maverick has been added since the 1982 *Catalogue of Cometary Orbits* was compiled. P/Hartley-IRAS (1983v), with an orbital period of twenty-one years, moves in an orbit tilted more than 96 degrees from the

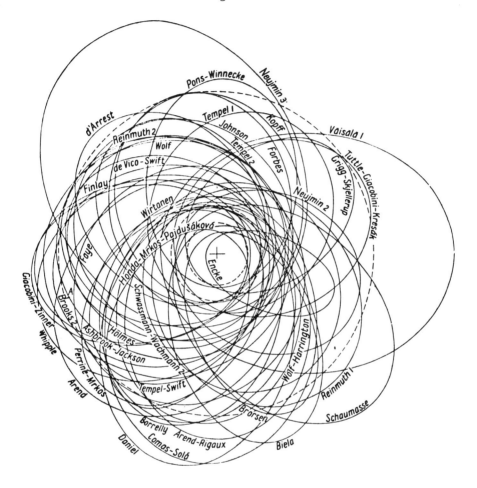

Orbits of typical short-period comets. Note dashed orbits of Jupiter near edge and orbit of Earth near center.

Earth's orbit and reaches some 14 AU at aphelion.

About a century ago, H. A. Newton at Yale University (not a descendant of our immortal bachelor, Sir Isaac) followed up a suggestion by Pierre-Simon de Laplace, whose explanation for Jupiter's comet family was based on the supposition that all comets were originally wild, coming from huge distances in space in nearly parabolic paths. What is the chance that Jupiter could catch them by its gravity and tame them into short-period, prograde orbits? He found that the chance is very small. Only about one in a million would have its period reduced to less than Jupiter's period of 11.86 years. The chance is still finite, however. Also he found that the probability for such capture is greatest for a prograde comet coming very close to Jupiter's orbit. That could account for the prograde motion of the tame comets.

Since the days of H. A. Newton, celestial mechanicians have wrestled with this problem and generally have agreed with him. Wild comets can only be tamed by means of Nature's usual wasteful technique: Lose a huge number, save a few. Because of the planetary attractions, about one-half of the nearly parabolic comets will immediately fly out of the Solar System, lost forever to outer space. The remainder will return, with their long periods somewhat reduced. Eventually, almost all will be lost. There must be a source to replenish them, some source of wild comets.

In searching for such a source, an Estonian, Ernst J. Öpik, then at Harvard, asked himself a related question: "What is the possible size of the swarm of comets that the Sun's gravity can hold?" If alone in space, the Sun could "keep" comets at almost infinite distances. But the Sun is not alone! Stars keep passing by to attract stragglers too far separated from the Sun. He concluded in 1932 that a cloud of comets in orbits out to 60,000 AU (1,500 times Pluto's distance) from the Sun are relatively safe from the attraction of passing stars for the life of the Solar System. In the 1930s the Earth's age was calculated to be only about 2 billion years, a value that is now well determined at about 4.6 billion years. Because of this, Öpik's dimensions for the cloud of comets must be slightly reduced today. His view of the Sun's comet cloud can be likened to a swarm of gnats, fired upon by a rifle. Each bullet eliminates a number of gnats as a passing star knocks out the comets near its path, but the swarm persists, only slightly diminished.

Öpik had to leave one question unanswered: Are most of the wild comets interstellar visitors, or are they all members of the Solar System? By 1950, Jan Oort of Leiden, in the Netherlands, could find an answer. Orbit computors had begun to trace back the motions of nearly parabolic and hyperbolic comets to determine their motions at great distances before they had come within the influence of the planets. Most had come into the Solar System in

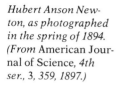

Hubert Anson New-
ton, as photographed
in the spring of 1894.
(From American Jour-
nal of Science, *4th*
ser., 3, 359, 1897.)

Ernst J. Öpik.

Jan Oort.

very long-period orbits, but some had been boosted into hyperbolic orbits by the planets. Still left over were a few whose original orbits were open, slightly hyperbolic. Oort believed that all the comets really belong to the Solar System and that the few discrepant orbits could be explained by the natural errors of measurement affecting calculations based on a limited number of observations.

From this point of departure, Oort developed a theory by which the passing stars can keep us supplied with comets. The passing stars simply knock a sufficient number of comets into the inner planetary region. Oort needed about a hundred billion comets in his cloud to do the trick. Later we will see whether Oort's requirements are reasonable.

Since Oort's time, the number of observed comets has increased by 50 percent, and the accuracy of the calculated orbits has improved greatly. The electronic computer makes possible calculations that previously would have taken many lifetimes to complete. Marsden and his colleagues find that, among 200 of the better observed comets with periods greater than 250 years, only 22 appear to have entered the Solar System from outer space, the others being bound to the Sun in long-period orbits. In no case does the excess velocity of the possible interstellar interlopers exceed the probable extreme errors made in the observations and in their interpretation. For example, the worst case is comet Sato (1976 I), which, according to calculations, had a velocity of 0.8 kilometers per second (0.5 miles per second) at infinity, before it headed into the Solar System. The Sun is moving among the nearby stars at a speed of some 20 kilometers per second (12 miles per second). If we are really seeing comets coming from the void, we should expect them to fly by much faster than just 0.8 kilometers per second.

Conclusion: With few exceptions, comets belong to the Sun's family, and are gravitationally attached. As an aside, note that, out of the 120 comets with periods longer than 61,000 years, 46 (or more than one-third) are calculated to be leaving the Solar System forever, as they are being kicked out by the gravitational slingshot effect of the great planets.

In the hands of Edgar Everhart at the University of Denver, the electronic computer gives us some idea of comet-taming statistics. His calculations follow 5,000 hypothetical comets for up to 1,000 round trips about the Sun. Of 100 lucky comets that start out in rather favorable orbits, coming in near Jupiter's orbit in prograde motion, less than 1 percent are captured into typical short-period orbits, according to Everhart's simulations. Even these few captures may be temporary, however, as Jupiter could nudge them out again after several revolutions. The overall chance for a wild comet starting in a favorable orbit to be captured into Jupiter's family is something like one in five thousand. The average comet probably must survive hundreds of revolutions to succeed.

Conclusion: The average short-period comet must have lived for hundreds of revolutions after having started out in very long-period orbits. To reverse our concept, and to suggest that short-period comets produce the long-period ones is quite impossible statistically, even if there were some means of making comets in the inner Solar System. Thus, we are dealing in the main with rather substantial,

D-G80-551-2

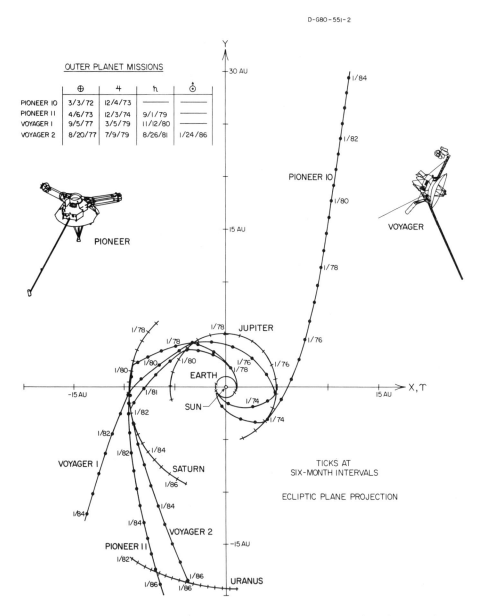

OUTER PLANET MISSIONS			
	⊕	♃	♄	⊙̇
PIONEER 10	3/3/72	12/4/73	—	
PIONEER 11	4/6/73	12/3/74	9/1/79	—
VOYAGER 1	9/5/77	3/5/79	11/12/80	—
VOYAGER 2	8/20/77	7/9/79	8/26/81	1/24/86

Diagram demonstrating how Jupiter and Saturn have been used as gravitational slingshots to change the orbits of four spacecraft and send them eventually to infinity. The dated points are for January of the year in question. (Courtesy James A. Van Allen and NASA.)

Typical short-period comet, P/d'Arrest. (July 31, 1976, by Elizabeth Roemer at Steward Observatory, University of Arizona.)

enduring entities, which are nothing like the evanescent upper atmospheric exhalations that Aristotle envisaged.

The large bag of comets that our comet hunters captured thus tells us a clear-cut story about the long lives of comets in the wilds of the Öpik-Oort cloud, their adventures with passing stars, and, finally, the domestication of some of them by the great planets. As always in science, we shall use all the clues we can find to test the validity of an attractive theory such as this one.

Chapter 8   **Some Comet Frailties and Idiosyncrasies**

Do comets really die? Our best evidence comes from several comets that have actually disappeared before our eyes. The classic example is comet Biela, which was discovered in 1772, not by Biela, but by Montaigne of Limoges, France, Messier's comet-hunting nemesis. The calculations of the comet's period were so inaccurate that its rediscovery by Pons in 1805 was not recognized as such for two decades. Only when an Austrian officer, W. von Biela, again found it in 1826, was its orbit of sufficient accuracy to associate it with the two earlier independent discoveries. Its orbital period was 6.6 years. Thus, Biela had the luck to have his name attached to a famous comet, his only discovery. Its predicted return in 1832 caused another great comet scare because calculations showed that it would almost "cut" the Earth's orbit. The newspapers never noticed that it would miss the Earth by 80,000,000 kilometers and cross the Earth's path a month before the Earth arrived.

   In 1839, Biela's comet was poorly placed for observation and could not be found. Its 1846 return brought a real surprise. The comet was double! Not only was it double, but sometimes one and sometimes the other component was brighter. At times, it had two and even three tails, with a highly variable nebulous coma around the two nuclei.

The two components of comet P/Biela, February 19, 1846 (by Otto Struve).

As might be expected, the next return of Biela's comet in 1852 was widely anticipated. By then, the two components were separated by about 2,400,000 kilometers and were moving in approximately the same orbital path. Again, sometimes one and sometimes the other component could be seen. Only the person computing the orbit could tell which was which. The 1859 return was unfavorable, but in 1866 no sign of either component could be found, in spite of excellent circumstances for observations. Biela's comet had split in two and disappeared, never to be seen again. Later its skeletal remains became evident, but that is a story for chapter 9.

Another disappearing comet was found by Justus Georg Westphal of Goettingen, Germany, in 1852. This comet, which could be seen with the naked eye, moved with an orbital period of sixty-one years, coming to perihelion just beyond the Earth's orbit. It was due to arrive again at perihelion on November 26, 1913. But comet Westphal barely made it. All seemed to be going well after its recovery on September 26, 1913. Yet by late October, its nucleus had disappeared, leaving only a large faint disk that faded to a shapeless glow. Three days before perihelion, the comet was barely visible, although it was well placed in the sky for observation. A few later uncertain observations were made in January 1914. At its next predicted return in 1976 no one could find it. Its fading in 1913, while on its way toward the Sun, when comets usually brighten rapidly, meant that comet Westphal was in its death throes.

Four other short-period comets have failed to reappear after having been recovered on at least one return and having been sought at several later predicted returns. A fifth one, P/Perrine-Mrkos, has been of interest for more than half a century. Discovered by Charles Dillon Perrine in 1896, this comet, which had a 6.4-year period, was rediscovered in 1909. Then it was lost for six passages. It flared up in brightness in 1955, enabling the skilled Czech astronomer Antonín Mrkos to spot it. Somewhat fainter in 1962, the comet faltered considerably in 1968. Soon after perihelion on November 1, it began to act like comet Westphal. It became increasingly diffuse and pale, finally disappearing from sight after January 17, 1969. In 1975 and 1983 no one could find any trace of P/Perrine-Mrkos.

Some comets in short-period orbits are cosmic mortals. Although most of them probably survive more revolutions on the average than humans live in years, their lifetimes must be trivial compared with the age of the Solar System, which is some 4.6 billion years.

Halley's comet has made twenty-nine revolutions since the Chinese observed it in 240 B.C. Has it changed much in brightness? David Hughes of Sheffield University, England, catalogued the various distances at which it was discovered and was lost to sight during its recorded appearances, thus deriving a rough measure of its changes in brightness. He found that it may have faded by a factor of about two in twenty-one centuries. But the uncertainty in his calculation is as great as the calculated amount of fading. P/Halley still stands out as a long-lived maverick revolving the "wrong way" around the Sun.

Some comets, although apparently rather sturdy under adverse situations, are by no means unaffected by them. Famous are the Kreutz sungrazing comets, first discussed thoroughly by Heinrich Kreutz of Germany in 1888. In 1882, the Great September Comet passed through the Sun's atmosphere at a distance from the solar surface of one-third the Sun's diameter. Before perihelion, the nucleus was seen to be single, but afterward the nucleus became greatly elongated with four to five condensations strung along it like beads. Barnard discovered six to eight comet-like objects nearby. The tail reached a length of 15 degrees, as seen with the naked eye. In spite of these vicissitudes in the several-thousand-degree temperature so near the Sun, the comet persisted 8.5 months after perihelion and was last seen at 4.4 AU from the Sun. Two of the fainter components lasted for only 5.5 months, a third only a few days longer.

The earlier sungrazer of 1843, coming even closer to the Sun—only 130,000 kilometers from the surface—and venturing deeper into the corona, was much more spectacular. It outshone the planet Venus by a hundred times. Even its tail became conspicuous in daylight. The longest tail ever observed, it stretched out in the late evening sky beyond the distance of the orbit of Mars, almost 2 AU in length. In spite of its daring adventure so close to the

The Great Comet of 1843 as seen from Blackheath Kent, on March 17. (Reprinted from G. F. Chambers, The Story of Comets *[Oxford, 1909], plate XIV).*

Sun's inferno, this remarkable comet failed to break up. Nevertheless, it remained visible for only about three months afterward, to a distance of 1.9 AU, fading away when much nearer the Sun than its 1882 relative.

The most spectacular sungrazer in recent times, comet Ikeya-Seki (1965 VIII), is named for the two Japanese amateur astronomers who discovered it. The comet, with a magnificent tail visible in the morning sky, became a thousand times brighter than Venus. The Sun's heat definitely went to its head, as the nucleus appeared to split in two. It faded from view after nearly four months.

No one knows exactly how many Kreutz sungrazers have actually been seen. Of the nine observed from 1668 to 1970, four have excellent orbits, with orbital periods ranging from 512 to 1,111 years, whereas five others could be fitted only with parabolic orbits because they were not well enough observed. All move in nearly the same path: tilted about 140 degrees, or retrograde, with respect to planetary motions. All are better observed from the Southern Hemisphere than from the Northern. One was probably seen in 1882 near the Sun during an eclipse. Other "eclipse" comets may well belong to the Kreutz group. Still others are suspected members, such as the one seen by the Chinese in 371 B.C.

*Comet Ikeya-Seki
1965 VIII, October 7,
1965, by Elizabeth
Roemer. (Official U.S.
Navy photograph.)*

*Comet Ikeya-Seki
1965 VIII with the
zodiacal light. Solar
system dust inside the
earth's orbit, illumi-
nated by sunlight on
November 1, as photo-
graphed by Alan
McClure, using 11-
minute exposure on
Royal-X Pan film with
a Bronica camera and
a wide-angle 50-milli-
meter lens.*

The Space Age brought a related cometary surprise in 1979. Three solar astronomers at the U.S. Naval Research Laboratory, R. A. Howard, M. J. Koomen, and D. I. Michels, were operating a satellite solar observatory (SOLWIND) when they discovered a comet deep in the Sun's corona. It seemed to be moving in a path characteristic of the Kreutz group, but it lost its will to live. It actually plunged into the Sun, leaving its tail behind, visible for a few hours thereafter. This was the first known case of a comet being immolated by the Sun. These solar scientists later found four other apparent comets very near the Sun; perhaps more will show up in their satellite records. Note that two completely different kinds of artificial satellites have discovered comets: a satellite for solar research sensing with visible light, and a satellite sensing with infrared light. These developments are typical of the current revolution in astronomy stemming from technological innovations.

Comet Ikeya-Seki, October 21, 1965. (Courtesy Nurikura Coronograph Station, Japan.)

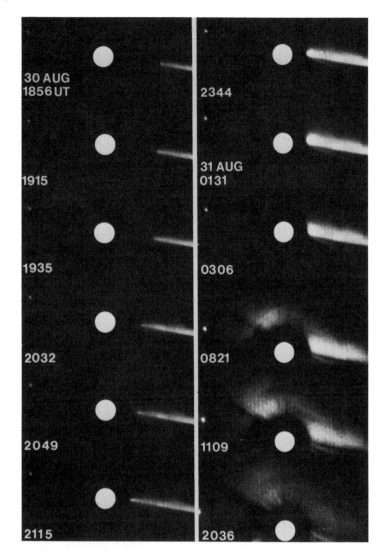

The comet that fell into the Sun, from August 30, 18ʰ 56ᵐ, to August 31, 20ʰ 36ᵐ, 1979. The dark occultation disk can be seen around the central simulated Sun. The comet enters from the right, leaving its tail temporarily behind. On August 31 at 8ʰ 21ᵐ, particles in the tail have been deflected by the Sun and can be seen on its upper left. (From SOLWIND satellite by R.A. Howard, M. J. Koomen, and D. I. Michels. Official U.S. Navy photograph.)

The sungrazers give us some important clues to the nature of comets. Some of them split when near the Sun, for example, and some do not. Many of them appear to survive, even though most of the split fragments may disappear rapidly. Their orbits supply additional evidence of past splitting episodes in the family history. The orbits of the Kreutz sungrazers fall into two groups, with slightly differing elements. It seems almost certain that a giant sungrazer once broke into many pieces, two or more of which, on subsequent returns, may have parented the two groups and the dozen or so uncertain members whose parentage remains in doubt. The fact that three were discovered within only a few months leaves us in doubt as to how many pieces of the parent comet or comets may still be in orbit.

The Kreutz group and Biela's comet are by no means the only comets that have split. At least twenty-one other comets, most of them long-period comets, have split, and the lesser components have persisted for anywhere from days up to several years. In recent decades, comet West of 1976 staged the best "split-tease" performance that a Space-Age audience had ever witnessed. In a very long-period orbit, this rather wild comet came to perihelion at a distance of some 29 million kilometers from the Sun. Soon after, without apparent cause, it split into four pieces. Two of the secondaries could be followed for nearly five months. The major component, lasting for another two months, disappeared from view at the respectable distance of some 3.7 AU from the Sun, after September 25, 1976.

Splitting does not appear to be related to position in the orbit or distance from the plane of the planets, where collisions with the multitudinous asteroids might occur. Only for the Kreutz group, and the short-period comet discovered by William R. Brooks in 1889, can we find an obvious reason for splitting. Whereas the Kreutz group nearly grazed the Sun, P/Brooks 2 went extremely close to Jupiter some three years before discovery. This suggests that gravitational tidal forces near the Sun or Jupiter caused these comets to split. Comet Brooks broke into four pieces. It still appears to be alive, though perhaps sickly. Having a seven-year orbit, it has been observed in

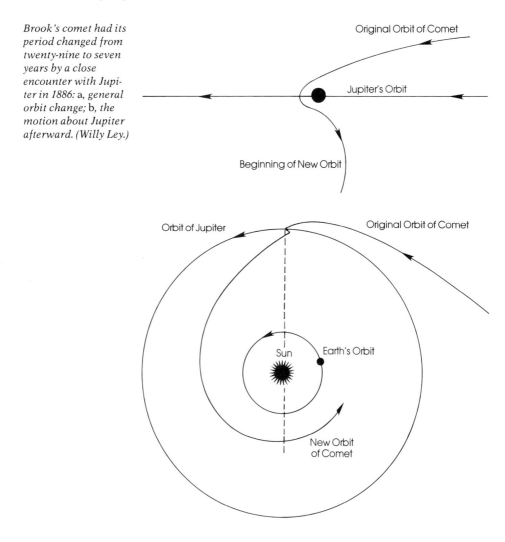

Brook's comet had its period changed from twenty-nine to seven years by a close encounter with Jupiter in 1886: a, general orbit change; b, the motion about Jupiter afterward. (Willy Ley.)

all but two of its thirteen solar round trips between 1889 and 1980. Recently, however, its brightness has fallen by a factor of about a thousand times from its value in 1889 and 1896. Senility may have set in.

So many comets have been known to split that the process probably takes place more than once in the life-time of an average comet and adds further testimony as to their mortal nature. Our problem, of course, is to deduce from these clues what actually splits in a comet when it comes too close to a massive body such as the Sun or Jupiter. And what splits in comets isolated in space? Why do they split at all? Before we can answer these questions, we must look at some other strangely behaving comets.

Many comets defy Newton's law of gravity! When well enough observed, most of them swerve away from their calculated motion by quite measurable amounts. We have already noted that Halley's comet disappointed the celestial mechanics by being tardy. The prime example is Encke's comet, named for Johann Franz Encke, a German astronomer who uncovered this quirk in the comet's motion. The story begins in Paris in 1786, when Pierre F. A. Mechain discovered a faint comet with the naked eye that he and Messier saw on only two nights. In 1795, Caroline Herschel, the devoted sister of Sir William (the discoverer of Uranus), found the same comet independently, but her few observations could not produce even a parabolic orbit. When it was rediscovered in 1805, the circumstances were similar to those of 1795. The great comet

Caroline Herschel, age 92.

hunter Pons found the comet again in 1818 and this time it was observed for seven weeks. When Encke calculated its orbit, he came up with a definite but amazing ellipse: one with a period of only 3½ years, the shortest period on record and, indeed, the shortest known to this day. Fascinated by the short period, Encke proceeded to look for previous discoveries and identified the comet with those of 1786, 1795, and 1805. The period came out to be 3.3 years! Seven appearances had been missed. This was a fine opportunity for Encke to test Newton's theory, including the effects of planetary attractions. He could also confirm his identifications and then predict the comet's next return. By extraordinary efforts, using only logarithms, Encke made the calculations in six weeks. He found that Jupiter would retard the 1822 return by nine days. His prediction was excellent. The comet of 1822, officially named P/Encke, is one of four, as we have noted, that have been named after someone other than its discoverer.

The comet returned on schedule five times up to 1838, during which interval Encke's calculations revealed a strange fact about the orbital period. It was becoming shorter—even though he had allowed for the attraction of the planets. Between 1789 and 1838 the period had decreased by 1.9 days, or more than 0.1 day each revolution! To explain the undeniable fact, Encke proposed that space contains some "thin etherial medium" that resists the motion of an extremely tenuous comet, but does not affect the motions of the massive planets. Encke's careful research is of greater interest today than his explanations.

We have already noted that Halley's comet came late in 1835 and in 1910. No interplanetary medium could both retard the motion of P/Encke and push forward P/Halley. The Prussian astronomer, Friedrich Wilhelm Bessel, well remembered today for having invented a new type of mathematical function, suggested prophetically that the reaction effect of material thrown out by a comet might change the comet's motion. But he never described a mechanism. By the 1860s, as we shall see, a new property of comets was discovered that buried Bessel's idea in the archives for more than a century.

Encke's comet as seen in 1838. After Amédée Guillemin.

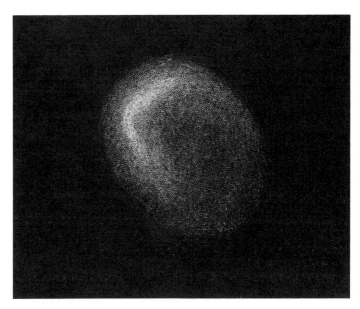

At this point in our story, we find that the comet puzzle has become even more intriguing. Comets defy Newton's law of gravity; some go too slow and some too fast. Comets tend to split in pieces, particularly when they are near the Sun or Jupiter, but also when they are quite undisturbed in space. Some comets seem to tire out and die. What clues do comets leave behind them along their orbits?

Chapter 9 Small Pieces of Comets

The *Georgia Courier* provided the following account of an event that took place on November 12, 1833.

At about nine p.m. the shooting stars arrested our attention, increasing in both number and brilliance until 30 minutes past 2 a.m., when one of the most splendid sights perhaps that mortal eyes have ever beheld, was opened to our astonished gaze. From the last mentioned hour until daylight the appearance of the heavens was awfully sublime. It would seem as if worlds upon worlds from the infinity of space were rushing like a whirlwind to our globe . . . and the stars descended like a snow fall to the earth. . . . Occasionally one would dart forward leaving a brilliant train which . . . would remain visible, some of them for nearly fifteen minutes.

According to some estimates, as many as 9,000 shooting stars, or *meteors*, could have been seen over the entire sky in the fifteen minutes following 5:45 a.m. This tremendous meteor shower frightened people from Missouri to the Atlantic Coast. It was even reported from ships at sea. Many awed spectators were certain that the Day of Judgment had, indeed, arrived.

More astute observers noticed that each meteor moved on a nearly straight path that plotted back to the constellation of Leo. Hence the meteors of the shower were called Leonids. We need not dwell on ancient explanations .of such showers, but should notice that a swarm or cloud of small solid bodies orbiting the Sun in space will strike the Earth's atmosphere in parallel paths and will become luminous on entering the high atmosphere at high velocity. At that point, however, their paths will appear to radiate from the direction opposite to their relative motion. One body traveling directly toward the observer will appear stationary as it moves precisely from this critical direction or *radiant* of the shower. All meteors start their visible paths when they strike the high atmosphere, a point that may appear quite far from the radiant to observers on the ground. All the trails project back roughly to the radiant point. The actual height of the Leonids in the Earth's atmosphere is somewhat greater

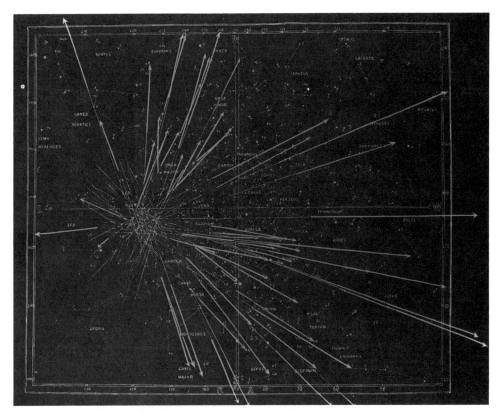

Leonid meteor shower
observed at Green-
wich, November 13,
1866.

The apparent divergence of parallel lines.

than 100 kilometers (60 miles) at the beginning of their trails.

Although the Leonid meteor shower that recurs near November 12 had long been a recognized annual event, it was of minor interest until the spectacular shower of 1833 attracted the attention of astronomers, who hoped to determine its true source. Some suspected that the similar great November showers in 1799 and 1698 were con-

The Leonid meteor shower of November 12, 1799, as it appeared to Andrew Ellicott at 3 a.m. off the coast of Florida. (Reprinted from Edwin Dankin, The Midnight Sky *[London]; courtesy Donald K. Yeomans.)*

nected with that of 1833. H. A. Newton decided that, if the Leonids represent a cosmic stream in orbit about the Sun, the dates should shift back in the calendar about one day in seventy years, to correspond to the motion of the Earth's equator with respect to the stars. The records showed thirteen strong showers back to A.D. 902 (on October 12). The dates, when corrected, indicated that the shower occurred near the same point in the Earth's orbit. In 1863, Newton published his paper showing the 33.3-year orbital period of the great Leonid showers.

Giovanni V. Schiaparelli, the Italian astronomer who described the "canali" on Mars, had also taken up the problem and published a book on it in 1866. He had not only noticed the period of 33.25 years for the Leonids, but had calculated the orbit for the meteor stream. The calculation of the six orbital elements is quite possible in this case because a point in the orbit is known, namely, the Earth's position when the stream strikes. This gives three quantities, the three-dimensional (x, y, z) location of the Earth in its orbit. The direction of the radiant gives two more observed quantities, which, with the assumed orbital period, add up to six quantities. These, with a few calculations, then provide the six orbital elements.

But Schiaparelli's book of 1866 contained another surprise. It reported that the orbit of the magnificent Swift-Tuttle comet of 1862 gave the correct radiant for the Perseid meteor shower, occurring regularly about August 12. At last, a comet was definitely associated with a meteor shower!

In 1867, within a few days of each other, Schiaparelli and two other well-known astronomers independently recognized that the Leonids were produced by the Tempel-Tuttle comet of 1866. Moving with nearly the orbital period of the comet (33.5 years), a cloud of small bodies near the comet's orbit crosses the Earth's orbit repetitiously, to produce an unusually strong meteor shower. On this basis, impressive Leonid showers were predicted in 1899 and 1933, but they turned out to be great disappointments. Therefore, expectations were not high in 1966, but this time the shower was strong. Jupiter's attraction seems to have deviated the major condensation away from and back to intersection with the

The Earth encounters a meteor stream.

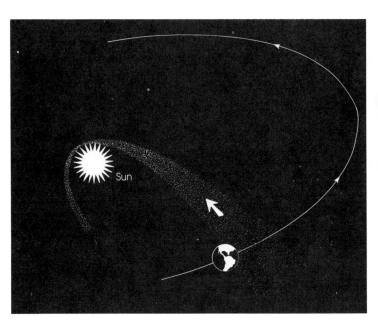

Earth's orbit. A few Leonid meteors are seen regularly during/between November 14 and 20—and are only a faint reminder of the magnificent displays of the past. No one knows whether they will ever be spectacular again. Their orbit is tilted about 18 degrees from the Earth's orbit, with motion in the opposite direction. Because of this, they strike the Earth head on, against its motion, at nearly the maximum speed possible, some 44 miles per second. At that speed, an ounce (28 grams) of meteor material has the energy of 30 pounds (14 kilograms) of TNT, which is released by friction with the atmosphere. An appreciable fraction of the energy appears as light. Even a pinhead mass in the Leonid shower can be seen from the ground as it streaks across the sky.

Thus two meteor showers, the Leonids and the Perseids, were known to have been caused by pieces of comets. The meteor bodies are clearly spread out along and near the orbits because of solar radiation effects. There is no need for the comet to be near the Earth to cause a meteor shower. Once the Leonid and Perseid showers were understood, astronomers quickly associated Biela's comet with a moderate shower that had been seen regularly from the constellation of Andromeda since 1772. The

Bright meteors photo-graphed by fixed tele-scope. (By J. S. Astapovitsch.)

breakup and disappearance of Biela's comet aroused suspicion that stronger showers might come. On the night of November 27, 1882, four observers in Italy counted 33,400 "Andromedids" in 6.5 hours. Meteors from P/Biela should emanate from the constellation Andromeda. An even more spectacular shower appeared in 1885, near the time that the comet might have returned. What better memorial to a dead comet?

Up to now, about fifteen comets have been identified as the sources of meteor showers. The orbit of Halley's comet misses the Earth's orbit by some 8,000,000 kilometers. Nevertheless, a few meteors from the comet are seen every year around May 8 (the Eta Aquarids), and around October 21 (the Orionids), near the times that the orbits of the Earth and Halley's comet come closest to each other.

An important clue in our story is the unique occurrence of meteors at the time of Encke's comet. Its orbit and the Earth's now miss each other by a wide margin. In spite of the lack of fit, I found in 1940, from double-station photographic meteor studies, that two Taurid meteor showers around November 1 came from orbits that are shaped like P/Encke's, but tilted from it in space. Jupiter is responsible; it causes P/Encke's orbit to swing around with a

Bright fireball in Oklahoma, January 3, 1970. It was photographed by two other stations simultaneously so that two freshly fallen meteorites could be recovered near Lost City, Oklahoma. (By Richard E. McCrosky, Smithsonian Astrophysical Observatory.)

period of some 5,800 years. The small rocks lost from P/Encke deviate from its orbit so that their orbits swing about at different rates; some finally intersect the Earth's orbit. This spreading out of orbits takes something like a thousand years or more. The important clue is that P/Encke must therefore have existed in more or less its present orbit for at least that length of time. I was happy to see my theory confirmed after World War II, when A. C. B. Lovell (now Sir Bernard) and his colleagues at Jodrell Bank, England, found daylight radio meteors in P/Encke-like orbits near the end of June, at the other point where tilted P/Encke orbits and Earth orbits can intersect.

Encke's comet may well be the Methuselah of comets, having survived passages within Mercury's orbit for many hundreds, if not several thousands, of revolutions. It has been seen at fifty-three apparitions (only one being missed since 1819) without the loss of more than a factor of two or so in brightness. Modern telescopic wizardry can now follow it around its entire orbit. Some comets are truly sturdy!

The great discovery in the 1860s that comets can disperse into meteor streams immediately led some to formulate a theory about their nature: They must be "flying sandbanks" or "gravelbanks" that lose their outer fringes to make the streams of small rocks or gravel that can strike the Earth's atmosphere as visible meteors. It was surmised that gas needed to make huge comet tails could have originated in the grains and been expelled by the solar heat. In the next chapter, we will learn more about this gas.

Once it had been established that comets are associated with meteor streams, the question immediately arose: Will we find samples of comets in meteoric particles? These particles consist of rocks and irons that have sur-

Great iron meteorite found near Willamette, Oregon. (Courtesy American Museum of Natural History, New York.)

vived the violent heating that takes place when they enter the atmosphere from outer space. The biblical and Chinese accounts tell of fiery and frightful stones falling from the heavens. In no case, however, even to this day, has a meteorite been seen to fall along a path corresponding to that of any well-known meteor shower. Comets clearly produce meteors, but none of the stones seem to be large enough or solid enough to be picked up as meteorites. Hence comets must contain sand or gravel with few, if any, large stones or irons among them. The corollary theory that comets are flying sandbanks or gravelbanks persisted for most of a century—a theory unchallenged because astronomers, beginning late in the 1800s became so enamored of the "rainbow" that most of them lost interest in comets.

Chapter 10　**The Magic of the Rainbow**

Sir Isaac Newton achieved another of his many scientific triumphs by demonstrating that white light actually is made up of all colors. He allowed sunlight into a dark room through a slit, then spread it out, by means of a prism, into a *spectrum* resembling the rainbow. With the proper optics, he could reverse the process so that the rainbow spectrum coalesced into white light again. As optical equipment improved, experimenters found that the appearance of spectra varied greatly from one source to another. In a flame, different chemicals show different bright sharp colors, from the violet through blue, green, yellow, and red. These are called *bright lines* when their spectrum is viewed with a slit set in front of the light source. It is important to note that each line has a unique color and, as we shall see, a unique wavelength. Hence we can think of a spectral line as a very sharp color or as a wavelength in the spectrum.

The Sun's spectrum was thought to be continuous, showing all the colors smoothly in sequence, until, with better optics, William Hyde Wollaston, an English physicist, found irregularities all across the colors (1802). About a decade later, Joseph Fraunhofer of Germany mapped out some 600 of these fine dark lines, or darker regions of the Sun's spectrum, known today as *Fraunhofer lines*.

A small section of the solar spectrum in the deep blue, showing the many faint absorption lines (most from iron).

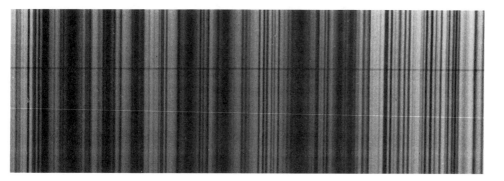

By 1859, Robert Kirchhoff at Heidelberg, could summarize the many spectral experiments in simple laws: When a substance, such as salt, is put in a flame, the spectrum shows bright lines—in this case, conspicuous yellow lines from the sodium in the salt. When a light from a continuous source passes through a cooler gas, the latter absorbs its own characteristic colors to produce dark lines. In either case, the gas is fingerprinted by its own peculiar lines, which are bright if hot gas radiates alone, or dark if seen in front of a hotter continuous source. Such a continuous source can be a hot solid, say, a poker, giving out all the colors of the rainbow. As it is heated, it becomes brighter and whiter, the blue light brightening more than the red. The Sun's outer atmosphere absorbs light (the Fraunhofer lines) from the hotter continuum below, the latter being at a temperature of some 5,500° C (9,900° F). All materials are gases in the solar inferno, but at high pressures a very hot gas acts like a solid and radiates a continuous spectrum. The gases in the cooler, outer solar atmosphere can be identified by their Fraunhofer lines, but only after physicists have studied the spectra of all

Kirchhoff's laws.

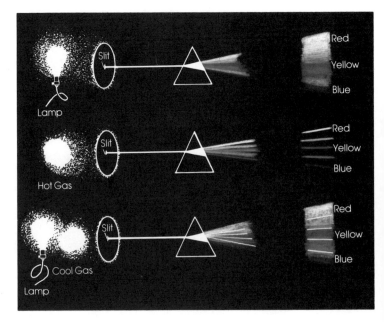

Solar corona during eclipse of November 12, 1966. Venus appears above. (By Gordon Newkirk.)

the possible substances in the laboratory.

Physicists identify the lines in the spectrum by measuring their wavelengths. Light moves at the incredible speed of 299,792.5 kilometers per second (186,284.4 miles per second). Each line (or color) has its own wavelength of vibration, which is roughly 0.00004 centimeters (1/60,000 inches) for violet light, and up to 0.000066 centimeters (1/40,000 inches) for red light. Besides wavelength and color, a spectral line can also be said to have a *frequency* of vibration equal to the velocity of light divided by the wavelength. A radio frequency can, in fact, be considered a line or narrow band of the radio spectrum.

With the advent of photography, the spectrograph (the optical device for making spectra) became a magic tool for physicists, chemists, and astronomers. At last the composition and temperature of celestial objects could be measured with confidence. The Sun turns out to be a cool star compared with some, but is typical of most. The Moon shines by reflected sunlight, so its spectrum is like the Sun's, even though the Moon itself is rather cold. Although almost all the known elements show dark lines in the solar spectrum, a huge number of the finer dark lines still remain to be identified. Helium announced its presence in the Sun before it had been isolated in the laboratory. In 1869, Charles A. Young of Princeton University located a bright green line in the spectrum of the Sun's corona when the disk of the Sun was eclipsed by the Moon. Because no such line was known in the laboratory, astronomers called the gas *coronium*. Seven decades passed before coronium could be identified, an indirect clue in the story of comets.

The spectrograph has another magic property. With it we can measure the speed of a celestial object traveling toward or away from the Earth. When the source of light approaches the observer, the waves are compressed or shortened, so the lines shift toward the blue, or shorter wavelengths. For a receding source, they shift toward the red, or longer wavelengths. Because light travels extremely fast, these shifts in the velocity of the spectral lines are minute and require extremely precise optics for accurate measurements. If, for example, a star is receding from us at a speed of 100 kilometers per second, the shift to the red in its spectrum is only 100/300,000, or about 0.033 percent.

Giambattista B. Donati of Florence, Italy, who discovered the great comet of 1858, was the first to observe a cometary spectrum—that of a comet discovered by Tempel in 1864. He saw three bright bands, wider than ordinary "lines." Pierre J. C. Janssen in France photographed the great comet of 1881 while Sir William Huggins in England and Henry Draper at Harvard later photographed its spectrum. Huggins found the sodium yellow lines, in addition to the three bands seen by

Spectrum of head and tail of Brook's comet of 1911. Note the continuous spectrum in the coma. (By F. Baldet, Meudon, France.)

Donati. He also saw some other bands, and, most important, the solar background light. His conclusion: Some solids must be reflecting sunlight. Thus began the physical study of comets.

Huggins set about to identify the gas that produced the bright emission bands in the comet's spectrum. He found that some of the bands coincided with "hydrocarbon" bands produced in the laboratory by spark discharges through such gases as ethylene (C_2H_4). The comet bands seen in the hydrocarbon spectrum actually are radiated by the double carbon molecule C_2. The violence of the spark breaks apart the hydrocarbon molecule so that the C_2 is separated from the hydrogen. These bands are called the Swan bands of carbon, in honor of the English physicist William Swan (1828–1914), a pioneer spectroscopist.

The violet bands in comets were later found to arise from a molecule of carbon and nitrogen, CN, which is quite unstable in the laboratory. The stable form is the deadly gas cyanogen, C_2N_2, so the cometary gas was also called cyanogen. This somewhat sloppy identification, not surprisingly, touched off another fearful rumor about

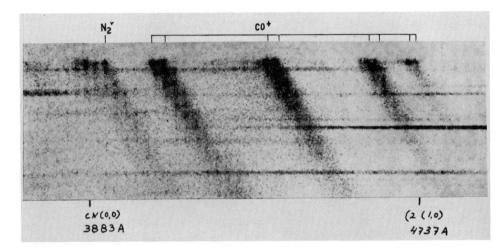

Strong tail spectrum of comet Morehouse in 1908. (By F. Baldet, Meudon, France.)

comets: Comets might poison us with lethal gas. In fact, many people failed to see Halley's comet in 1910 because they locked themselves in their rooms to avoid being gassed when the Earth passed by the edge of the comet's tail. Hawkers made brisk sales of "comet pills" and gas masks to protect the comet watchers. In Texas, people actually prevented the police from jailing the pill salesmen in the belief that the useless pills could circumvent the cometary hazard.

The brilliant sungrazing comet of 1882 developed bright metallic lines, particularly of iron, because the comet came so close to the sun that the heat could vaporize refractory metals. The comet's temperature rose to some 1,900° C (3,000° F). The comet's rapid motion deflected the lines toward the red end of the spectrum, corresponding to some 26 kilometers per second, with respect to the Earth, which was close to the calculated speed of 28 kilometers per second. Thus comets, in the very early days of astronomical spectroscopy, showed bright gas bands of carbon and nitrogen, direct sunlight reflected from solids, and the shifts in wavelengths expected from their rapid motion through space.

By 1900, another unstable molecule of carbon and hydrogen, CH, was also identified and added to the list of clues, but it did not reveal much more than was already known about the true nature of comets. The great breakthroughs in physics were still to come. Were comets self-luminous, perhaps because electrical discharges produced the bright lines? Or was sunlight alone the energizing agent? If so, what was the unknown mechanism? How much gas was being produced? Why were some of the gases so unstable? No chemist can keep C_2, CN, or CH in a bottle. The study of the cometary rainbow produced more questions than answers. Meanwhile, the telescope and photography were bringing in a host of clues about the activity in the heads and tails of comets.

Chapter 11   **The Heads and Tails of Comets**

When Friedrich Wilhelm Bessel theorized that comets were ejecting something toward the Sun, he had a sound basis for the idea. Bessel was not only a renowned mathematician with deep physical intuition; he was also a keen observer. Halley's comet in 1835 gave him the opportunity to study closely the fine structure in the head of this nearby brilliant comet. His drawings, which are still of great value today, show jets, rays, and fans that appear to be leaving the *nucleus* in the solar direction. Many of these inner structures seem to bend or curl back, as if pushed by some repulsive force from the Sun. John Herschel, the son of Sir William, saw the same structures and motions from the Cape of Good Hope. Within an interval of only a few hours, the inner region of the *coma* or head of the comet changed markedly. Some great activ-

Friedrich Wilhelm Bessel. (Courtesy Sky and Telescope.*)*

ity was surely taking place, but it was difficult to explain. Were electrical charges on fine particles in the comet opposite in sign to the charges on the Sun? If so, electrical repulsion could force the jets of particles away in geometric forms, more or less like parabolas.

Using this assumption, Bessel assembled his mathematical tools and calculated the curves that the particles should follow. He tried various laws for the repulsion and coped with the difficult problem of the geometry. As always with a comet, we look at its projection on the sky. All the geometry is known: the direction to the comet, the direction to the Sun, the distances, and the projection of the orbit in space. What confronts us is a three-dimensional puzzle. To try to visualize the true structure from its projection on the sky gives one a headache.

Some of the accompanying drawings of Halley's comet and Bessel's theoretical curves give us an inkling of the complications that Bessel happily managed to put into physical and mathematical form. The basic idea, fortunately, is simple. Deep in the center of the coma at the nucleus is some type of multiple "fountain." Most of the material, particles, or whatever we see, spouts out toward the sun. Just as the spray from a water fountain curves back to the ground because of gravity, so the Sun pushes away the material in the comet, and the result is much the same array of forms. In a fountain, one can actually simulate what Bessel saw in comet Halley. Choose a bright,

Bredichin's classical fountain concept of sunlight pressing away the outgoing material from head of a comet.

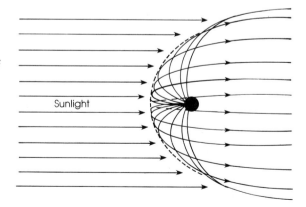

Sunlight

sunny day and stand near a typical water fountain on the Sun side. Spread the fingers of one hand and move the hand up and down in front of your eyes like a multiple shutter. Depending on the speed of your hand, you will see either water droplets or short traces of their trajectories. Now try to record what you have seen by setting your camera at different exposure times.

Comet Donati in 1858 brilliantly illustrated Bessel's fountain concept. The accompanying drawings by George P. Bond with the "great" 15-inch refracting telescope at the Harvard Observatory show the ideal fountain outline

Donati's great comet of 1858. Top, *October 4;* bottom, *October 5. (Drawings by George P. Bond, Harvard College Observatory.)*

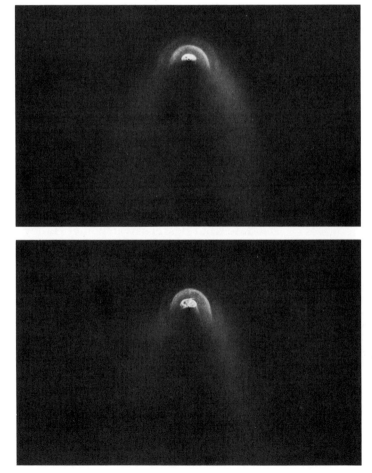

around the head of Donati's comet. Dozens of observers in Europe drew almost identical, but usually less artistic, pictures of the comet's coma—Bond was assisted by artist James W. Watts of Boston in preparing his published prints. His original notes show the same details in rough pencil sketches. Donati's comet, one of the most beautiful comets of all time, kept throwing off fountainlike envelopes regularly for several weeks. It was only a partly tamed comet with an orbital period of some 2,000 years, moving in an orbit tilted up 116 degrees to the plane of the Earth's orbit.

Coggia's handsome comet in 1874 showed the fountain envelopes as clearly as Donati's had, even to the repetitious ejection of envelope within envelope. It was a somewhat wilder comet than Donati's, moving with an orbital period of some 14,000 years, in a highly tilted orbit of 66 degrees. Many other comets confirm the fountain structure, but these two were brilliant and especially well placed for observing.

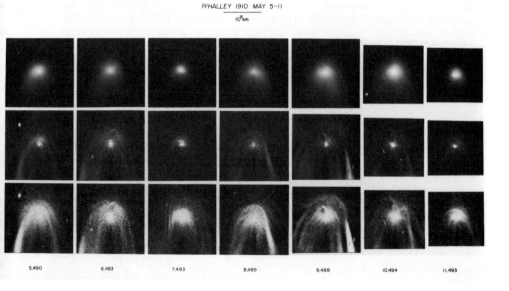

P/HALLEY 1910 MAY 5-11

10⁵km

5.490   6.483   7.493   8.489   9.488   10.494   11.493

Nuclear region of Halley's comet May 1910 at dates indicated. Width of frames: 150,000 kilometers. (Mt. Wilson photographs processed by Steven Larson to show details of near-nucleus structure.)

One of the charms of studying comets is their intriguing variety. Many comets show nearly round heads, or comas; some have a sharp stellar nucleus; some a central condensation; and some only a patchy, hazy coma. The comas range in diameter from tens to hundreds of thousands of kilometers. If nearby, the comet may appear as a fuzzy patch that looks as much as six times larger than the size of the Moon's disk. Such was the case for the comet to come nearest to the Earth in recent history, IRAS-Araki-Alcock, 1983 VII. Many comet comas are asymmetric or skewed, and connect with a wide variety of tails. The fountain effect shows up best in a bright comet when the geometry is right—that is, when the comet gives us a side view, as in the case of P/Halley in 1835, comet Donati in 1858, and comet Coggia in 1874.

Bessel's fountain model immediately explained the geometry of the tails. In space there is nothing to stop the flow the way that the ground stops a curtain of water in an earthly fountain. The fountain streams in space can move freely in the vacuum over the huge distances observed—in one case, more than 300 million kilometers. But what turns on the spigots of the fountain? And what turns them off? We need many more comet clues to answer such questions.

The tails, of course, are the spectacular parts of brilliant comets. Note how they differ from one comet to another. The Chinese called them "brooms" (flexible Chinese brooms, of course, not Western brooms). Other descriptive terms were bearded, sword, scimitar, and handle. Before the science of comets could progress, however, the tails had to be classified. About a century ago the Russian astronomer, Feodor A. Bredichin, set about the task. From all the observations he could collect for some forty-one comets, Bredichin derived three classes of tails.

Type I tails are nearly straight. They appear to curve back slightly from the motion of the comet when the geometry is right, that is, when the Earth, carrying the observer, is well out of the comet's orbital plane. Type I tails are often quite complex in form and may change rapidly from day to day, or even hour to hour. Knots, kinks, and irregularities usually break the even structure over dimensions of many thousands of kilometers in these

Comet Ikeya 1963 I on February 15, 1963. Example of a purely gaseous comet. (Smithsonian Astrophysical Observatory, Woomura, Australia.)

huge appendages. The Type I class comprises many of the most spectacular tails because these stretch out in a direction nearly opposite to that of the Sun. Observers in Bombay, for example, could see the tail of the great sun-grazer, comet 1843 I, extending from the horizon to the zenith. We have already noted its prize-winning length, which was in excess of 300,000,000 kilometers. Only a truly colossal solar force could produce such a fountain.

Bredichin's Type II tails fall in between the other two types and are not always clearly separated from Type III. Today's astronomers generally lump Type II and Type III tails together as Type II. Bredichin's Type III tails are short, stubby, and highly curved, again streaming behind from the orbital motion. They usually show little structure, their form changes slowly, and they seldom stretch out for more than 10 million kilometers. Most *bright* comets exhibit both Types I and II (or III) tails simultaneously, although a few "nonconformists" may have only one type. Seldom, however, is there no trace of a Type I tail. For faint comets that show any tail at all, the type is not always discernible because faint tails are usually rather short. The geometry, too, may disguise the type of tail, particularly when the Earth is near the orbital plane of the comet.

Bredichin's three types of comet tails.

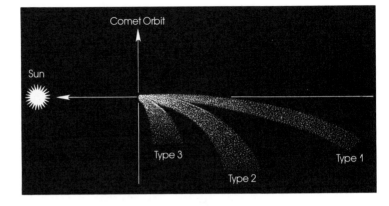

Extending Bessel's fountain theory, Bredichin suggested that Type I tails are made of the lightest atoms, hydrogen, blown away at the fastest rate by solar electrical repulsion; Type III tails are made of the heaviest observed atoms, iron; and Type II tails represent the molecules of intermediate weight, the hydrocarbons. Bredichin followed the physics of his time and the early spectral conclusions about the composition of comets. We shall find that this explanation is logically and intuitively brilliant, but, unfortunately, incorrect in its final conclusions.

The curvature of the tails is explained by the Bessel-Bredichin theory, including the solar-repulsion theory and the known physical laws of motion. A particle, atom, or molecule ejected in a vacuum and forced away from the Sun must move in a hyperbolic path and lag behind the comet's motion in its orbit. The greater the repulsive force, the straighter the path in which the particle moves away from the comet. If the repulsive force is small, the particle moves slowly, lagging behind the comet to give a highly curved, but small, tail. Observations clearly show that all the tails stay very close to the plane of the comet's orbit; this is consistent with the assumption that the forces radiate from the Sun.

Bredichin noticed some intriguing motions in the long, Type I tails. The kinks and knots moved rapidly away from the head of the comet and from the Sun at speeds of several kilometers per second near the head. But their speed increased farther out, showing that the mysterious

Comet Mrkos 1957 V showing extremely strong tails of Bredichin Types II and III, as well as a fainter tail of Type I. (Courtesy Alan McClure.)

Rapid changes in the tail of comet Whipple-Fedtke-Tevzadze 1943 I over a period of 1.8 hours on March 29, 1943. (After C. Hoffmeister; courtesy Sonneberg Observatory.)

force is many times greater than the Sun's force of gravity. No satisfactory explanation for these strange forces was found before the 1900s. Electrical repulsion did not fit well theoretically. Two other ideas were talked about: light pressure from sunlight, and streams of tiny particles that shot out from the Sun. Although James Clerk Maxwell's theory of electromagnetism predicted that light should exert pressure, no laboratory measures had yet shown that light could really do so. But flares on the Sun and sunspot activity seemed to be the cause of displays of aurora borealis and magnetic storms on the Earth. As early as 1892, the Irish physicist George F. Fitzgerald calculated that these delay times from Sun to Earth gave a speed of about 450 kilometers per second for whatever kinds of particles were being ejected from the Sun. Barnard suggested in 1909 that the Sun in some way produced an effect in comet tails and produced aurorae by the same mechanism. What was really happening? The answer to this question had to await the arrival of modern physics.

Let us now return to the obvious center of comet activity, the nucleus buried in the coma. This elusive nucleus had become more and more of a mystery, even in the flying sandbank model. Some comets show only hazy centers or condensations in their comas, whereas others sometimes show sharp stellar centers. Even for comets that have come very close to the Earth, the best telescopes have not been able to resolve any "true nucleus" less than several tens of kilometers in diameter, in most cases, a few hundred kilometers in diameter. When Lexell's comet came only 2.4 million kilometers from the Earth in 1770, for example, the most highly trained observers could see no sharp center less than 100,000 kilometers across. The one exception was P/Pons-Winnecke, which passed within 6 million kilometers of the Earth in 1927. Its sharp, stellar nucleus could not be resolved as a disk, and so the diameter of the nucleus was certainly less than 5 kilometers. Similarly, no comet traveling directly between the Earth and the Sun—such as Halley's comet in 1910 and the brilliant sungrazing comet 1882 II—has been seen crossing the Sun's disk. The lack of observed shadows on the Sun's

Comet Seki-Lines 1962 III. (Photographed by Alan McClure on April 23.)

Comet Bennett 1970 II, with a dust tail more than 20° long, as photo-graphed by Alan McClure on April 5, 1970. Ten-minute exposure with a 105-millimeter lens and a blue-sensitive plate.

disk indicates that the diameters of these two comets are less than 100 and 70 kilometers, respectively.

Furthermore, comets seemed to have no appreciable mass because the Earth, for example, has paid not the slightest attention to the nearby comet, in the sense of gravitational deviation in motion. The same was true for Jupiter's moons, when Brooks' comet wandered among them in 1886. But Jupiter changed the comet's period from twenty-nine to seven years and caused it to split! Comets therefore could not "weigh" more than about a millionth of the Earth. Clearly, comet heads and tails must be great bags of "nothing." The meteor streams associated with comets add up to a negligible mass. Stars could be seen shining through comets without any measurable dimming, except perhaps in a few cases close to the nucleus—and even these near-nucleus observations were in question. How much "nothing" is there in comets?

We have learned that comets are like fountains, ejecting material toward the Sun, which, in turn, forces the material away to form the tails, great or small. To understand these forces and processes, to find numerical values for the masses involved, and to answer many other questions about comets, the astronomer had to turn to the physicist. Note that the first great breakthrough in physics had been made by astronomy in the seventeenth and eighteenth centuries, and that it gave real and colossal dimensions to the Universe and provided the fundamental law

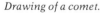

Drawing of a comet.

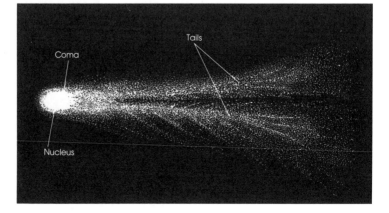

of gravity. Even the speed of light, the most fundamental of all physical constants, had first been measured by an astronomer, Olaus Roemer of Denmark in 1675. In timing the eclipses of the satellites of Jupiter, Roemer found a rather accurate value for the speed of light. During the nineteenth century, the tide turned. Astronomy became a branch of applied physics. The era of electrons, X rays, radioactivity, and tiny, quantized packets of energy was opening up. What has the new physics told us about comets?

Chapter 12 **Sunlight and Comets' Tails**

As mentioned earlier, the eminent Scottish physicist James Clerk Maxwell showed theoretically that light should exert pressure. Some three decades later, in 1900, Peter N. Lebedew in Russia, and E. F. Nichols and G. F. Hull in America, demonstrated the phenomenon in their laboratories. At last, the motions in great comet tails could begin to make sense. In 1908, comet Morehouse produced marvelous images on the newly improved, more sensitive photographic plates; its head and tail provided a celestial "laboratory" for testing the Bessel-Bredichin fountain theory. A. S. Eddington, the famous astronomer, later known as Sir Arthur, measured the parabolic envelopes and the motions of the knots and kinks in the huge tail of comet Morehouse. He was dismayed by the magnitude of the solar repulsive forces that he found there— they were close to 800 times solar gravity. Halley's comet in 1910 showed similar forces in the tail, averaging about 200 times the force of gravity, but sometimes reaching 1,000 times! Our assiduous comet hunter, E. E. Barnard, observed even stranger effects in the tail of comet Morehouse. Some areas brightened up without any visible source of material.

Almost everyone seemed satisfied that light pressure was the force, even though physicists could not account for forces greater than perhaps 200 times solar gravity, either in the laboratory or in theory. Astronomers, as we have seen, had become so entranced by the "rainbow"— that is, with spectroscopy of the Sun, stars, and nebulae— that they tended to neglect comets after P/Halley's stunning display in 1910. The puzzling problems of comet tails failed to arouse their interest.

Early in the twentieth century, progress in the physics of matter and light was breathtaking. Atoms were no longer tiny, hard balls bouncing against each other in gases and liquids and somehow tied together in solids. By 1913, the Danish physicist Niels Bohr had visualized them as sub-sub-microscopic "solar systems" with heavy, posi-

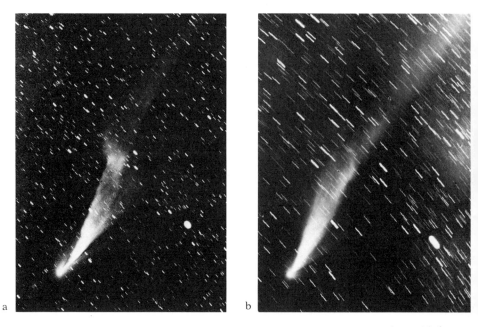

Comet Morehouse 1908 III, on September 30, 1908. Note the rapid changes in the tail between a and b, which are separated by 3ʰ 41ᵐ. (Photographed by Edward E. Barnard; courtesy Yerkes Observatory.)

tively charged nuclei attracting light, negatively charged electrons in orbits around them. The positive charge on the nucleus matched the sum of the negative charges on the orbiting electrons, leaving the whole atom neutral, that is, with zero charge. Thus the hydrogen nucleus carries one positive charge to match its single electron. Helium carries two charges for two electrons; carbon, six; nitrogen, seven; oxygen, eight; on to uranium, which carries ninety-two. The atom turned out to be as incredibly small as astronomical distances are incredibly large. The atom of hydrogen, for example, is only about one hundred millionth of a centimeter in diameter. If all the hydrogen atoms in a cup of tea were beaded together on a string, the string would stretch out to fifty times Pluto's distance from the Sun! Each atom is exactly like the others.

But this picture of an atom as a small solar system became outmoded in about a decade, even though the dimensions remained unchanged, as did the broad concept of positively charged nuclei and negatively charged electrons. The latter became statistical probabilities, dissolving into a haze around the positive nuclei, after the development of the new theories of *wave mechanics* or *quantum mechanics*. These abstruse mathematical concepts of atomic structure were developed by a new breed of physicists such as Louis de Broglie in France, Erwin Shrödinger and Werner Heisenberg in Germany, Paul Dirac in England, and others. Their mathematical models fit a host of laboratory measures, but describe atoms that are really like nothing we can see, feel, or experience. Physicists can only visualize the internal structure of atoms by means of equations representing highly tuned vibrators, held together by strange forces.

Light also became too complex to be visualized in terms

Albert Einstein at Harvard College Observatory in 1935. (By Margaret Harwood.)

of sensory experience. As Newton supposed, light *does* behave like a wave. It has a fixed speed and each color has a measurable wavelength. But Max Planck of Germany showed in 1900 that light also comes in little packets of energy, called *quanta*. The energy of each quantum is proportional to a universal constant divided by the wavelength. These quanta carry all types of electromagnetic radiation, from radio waves many kilometers in length, down through heat or infrared radiation, to red, yellow, green, blue, violet, and ultraviolet light, and on to the X rays and gamma rays of radioactive atoms. The range in wavelength and in the inversely related energy of quanta defies imagination. Whereas the human eye detects light over about one octave or a factor of two in wavelength, the heat in the infrared spectrum covers eleven octaves, while radio covers twenty-seven octaves, if we stop at radio waves 100 kilometers in length. This adds up to thirty-eight octaves, or a factor of 300 billion times in wavelength as we go to lower and lower energy quanta. In the other direction of higher energy quanta and shorter wavelengths, ultraviolet covers five octaves, X rays about ten, and gamma rays at least ten more octaves. Hence, the total observed electromagnetic spectra cover at least sixty-four octaves, or a range of 10^{19} times in energy. The gamma-ray quanta are most energetic and may carry a billion-billion times the energy of long radio waves.

All these radiations travel at the same speed—that of light—measured by astronomers to be constant to one part in a trillion over a wavelength range also of a trillion. But all these forms of radiation over this enormous range of energy and wavelength obey the same laws of quantum mechanics! This quantum picture of light or radiation was shocking enough in concept, but Albert Einstein added another shock by asserting that mass and energy are actually interchangeable, being related by the amazingly simple equation: Energy equals mass times the square of the velocity of light ($E = mc^2$).

For comets, we need only to understand (vaguely) how light reacts with atoms and molecules. At the moment, we need discuss only atoms in a low-pressure gas, because molecules react to light quanta in essentially the same way as atoms. Remember that a quantum corresponds to a line in a spectrum. An atom by itself has a complicated internal vibrational system that settles down to a resting, or "ground state," of energy. This system simply "sits" quietly, doing nothing unless a quantum energizes it or another particle collides with it vigorously. Internally, the atom is tuned to respond to light quanta of certain highly specific energies. We may liken it to a violin string at rest, or to a crystal glass. These will resonate when a sound wave of a certain pitch or pitches comes along. The violin string will vibrate and the glass may even shatter if the sound of the correct pitch is strong enough. The atom, like the glass, may shatter if it swallows too energetic a quantum; that is, it may become ionized, losing one or more of its electrons. Otherwise the atom, on swallowing a quantum, will quickly radiate away the energy it absorbs. Because it has many internal vibration levels, or energy levels, it may or may not reradiate exactly on the same wavelength as the quantum that excited it. It will, however, radiate away precisely the total energy that the quantum provided, although the energy may be divided up into a number of quanta or wavelengths. Hence, we may see a number of lines in the spectrum from atoms that have absorbed quanta of the same energy or wavelength. These lines will all be of longer wavelength (redder) than the original line corresponding to the exciting quantum.

Almost all the lines we measure in the visual spectra of comets are reradiation from ultraviolet light that the atoms or molecules have absorbed from sunlight. The atom or molecule "chooses" its mode of radiation according to statistical laws that have been developed by many physicists from laboratory measurements and related theory over several decades. Their task of measuring these energy levels, and of determining the probabilities of absorption and reradiation, is essentially completed for atoms but is unending for molecules, which provide an almost infinite variety of atom combinations. The atoms in a molecule are held together by forces derived from their electron systems. These forces allow vibrations to take place between atoms and also enable atoms to rotate around each other. As a result, even the simple cometary molecule CN (so-called cyanogen) shows hundreds of lines. In molecules, the rotational energy changes are small compared with the vibrational energy changes. A vibrational level may be associated with a host of nearby rotational energy levels. Hence, many rotational lines are bunched closely together in the spectrum, corresponding to many rotational energy changes for one change in vibrational energy. These bunched lines, which characterize spectra from molecules, are called bands.

We can now apply this long tutorial on light to some unanswered questions about comets. The dust or solid particles simply reflect sunlight. How they do it, as we shall see, tells us a great deal about the dust. The gases, however, are excited energetically as they absorb sunlight, usually in the ultraviolet. They then radiate away this absorbed energy in lines or bands that we can observe, usually at longer wavelengths, or redder than the lines that they absorb. This process is known as *fluorescence*. The mystery of comet light is now "solved": It is both the reflection of sunlight from dust and the fluorescence of sunlight from gases. Comets, like the planets, would be quite invisible without the Sun's radiation.

Atoms in sunlight are vulnerable. If they swallow a light quantum that is too energetic, such as one in the far-ultraviolet, they may suffer the loss of an electron and become ionized. They will then become positively charged and their entire inner structure will change. These atoms can absorb or radiate entirely new sets of lines and can be identified as ions from their spectra. Hydrogen, having only one electron, is quite helpless when ionized. Its positive nucleus, a proton, is optically dead in deep space, and cannot be detected unless it encounters an electron, an ion, or an atom. So is the nucleus of the helium atom after it has lost its two electrons and becomes an *alpha* particle, which has a double-positive charge and nearly four times the mass of a proton.

Molecules are even more vulnerable to sunlight than are atoms. Not only can molecules be ionized, they can also be broken apart, or can be *dissociated*, by energetic quanta. Thus our cyanogen molecule CN lives for only about a day at 1 AU from the Sun before it is split apart. Improved observations of comets, laboratory studies of

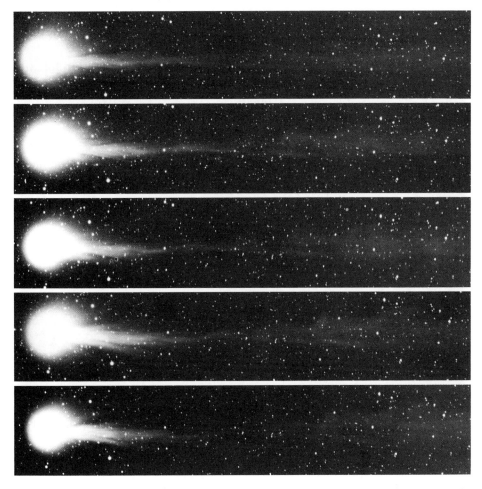

Comet Whipple-Fedtke-Tevzadze 1943 I, over a period of 6ʰ 37ᵐ on March 23 and 24, 1943, showing rapid motion and partial separation in the tail. (After C. Hoffmeister; courtesy Sonneberg Observatory.)

spectra, and theoretical developments combine to explain the nature of the long, Type I tails. These tails shine entirely by the fluorescence of ionized atoms. Type I tails are really ion tails.

Now we can summarize in a few words the surprising story that spectroscopy tells us about comet tails. The curved Type II tails shine because sunlight is reflected from solid grains. The physics of light pressure, which pushes these particles away from the Sun, will enable us to deduce more about their nature.

When we look at Type I tails, we are seeing solely the light from ionized atoms or molecules, those that have lost an electron and are therefore positively charged. This means that the Sun selectively repulses *ions*, not neutral atoms or molecules, with fantastic forces of up to a thousand times solar gravity! By midcentury the ions of CH^+, OH^+, CO^+, and nitrogen (N_2^+) had been identified in the great ion tails. How can the Sun accomplish this formidable task?

Chapter 13 The Prodigal Sun

In 1951, the astronomer Ludwig Biermann of Germany fitted together more of the many pieces of the jigsaw puzzle to explain the remarkable motions in the ion tails of comets. He knew that the Sun was spewing out gases that propelled ions but not neutral atoms or molecules. The glorious corona surrounding the Sun during eclipses provided another clue. The mysterious bright green line of so-called coronium had defied identification from 1869 until 1940. Bengt Edlén of Sweden finally discovered that coronium is not a new exotic element, just commonplace iron heated to about a million degrees at very low pressures. The iron atoms have lost thirteen of their twenty-six electrons. The light quanta at this temperature include not just ultraviolet light, but also soft X rays. Hydrogen at such temperatures, as we have seen, is impotent as a radiator, because it consists only of a proton; so is helium, with its two electrons torn away. Even though

Ludwig Biermann.

the corona consists mainly of hydrogen and helium, the light we can see comes primarily from heavier atoms, such as iron, nickel, and calcium, which can still hold onto some of their electrons after being tortured in a near vacuum at a million degrees temperature. Violent "bubbles" of turbulent gas and magnetic fields rise through the surface of the Sun to activate the outermost atmosphere of the Sun—its corona.

No wonder, then, that the solar flares can shoot off streams of ionized gas that impinge on the Earth's high atmosphere and produce the beautiful aurora borealis and unwelcome magnetic storms on the Earth. Early in this century, K. Birkeland of Norway developed a theory to explain these geophysical phenomena, and actually simulated them with a miniature magnetic Earth in his laboratory; no one then could picture the hellish violence in the majestic solar corona. Birkeland's theory was largely neglected because the solar link was missing.

Biermann could argue that the Sun *continuously* ejects a million-degree *plasma* of ionized gas, not just at the times of solar flares or unusual sunspot activity. The *solar wind*, blowing at 400 kilometers per second is not sporadic, but is more like a trade wind that occasionally resembles a hurricane. With the solar wind *always* blowing, every comet with gases ionized by sunlight can produce an ion tail. The ions are caught by the solar wind and carried away from the comet and the Sun at high speeds. The neutral atoms and molecules hardly notice the solar wind, although some are ionized as they jostle with the hot electrons in the plasma. The new ions then join the others to be swept into the ion tails.

In 1957, the Swedish physicist and Nobel Prize winner, Hannes Alfvén, applied his original theory of hot plasmas to comet tails. This subtle theory describes how magnetic fields can be carried by the hot plasma gases from the solar atmosphere into open space with the solar wind. Physicists and astronomers like to express the idea by saying that the magnetic fields are "frozen" into the hot plasma (at a million degrees temperature!). The magnetic fields in the solar wind act like fish nets and catch the ions near a comet to produce the great ion tails. Alfvén's theory has great bearing on all hot plasmas in astron-

*Aurora as photo-
graphed from College,
Alaska, on April 9,
1957. (Courtesy V. P.
Hessler.)*

*A huge solar promi-
nence on August 21,
1973, from Skylab;
imaged in far-ultravi-
olet helium light.
(Courtesy U.S. Naval
Research Laboratory.)*

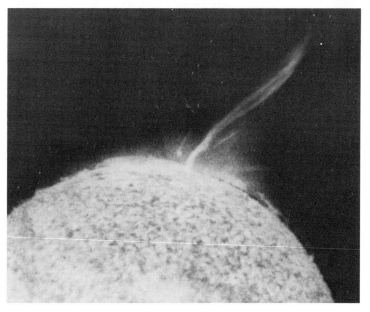

Solar activity on June 10, 1973, from Skylab; imaged in far-ultraviolet light. Left, in helium; right, in iron ionized thirteen times.

omy—stars, black holes, and even interstellar gases—but the theory is almost unmanageable in practice. It is a challenge to theorists; it must be mastered, however, if further progress is to be made in understanding the universe.

Biermann's theory was beautifully confirmed with the help of the early satellites of the Space Age. In 1958, the Iowa physicist James A. Van Allen discovered magnetic belts around the Earth, subsequently named for him. A particle detector on the first U.S. satellite, *Explorer I*, became overloaded by the unexpected violent flux of high-energy ions trapped in the Earth's magnetic field. Later satellites measured the solar wind directly. The prodigal Sun blows off a million tons per second of its substance in the form of this million-degree ionized gas. Wasteful as this may seem, astronomers find the Sun is relatively conservative when compared with many stars; some newer or brighter stars blow out a thousand times as much gas plasma as the Sun every second. The cause is now somewhat understood. Violent turbulence well beneath the visible surface of the Sun and such stars produces shock waves, which carry energy through the outer layers to be dissipated as a *stellar wind*. Internal magnetic fields play a role in the process.

*X-ray image of the
Sun on June 1, 1973,
from NASA's Skylab.*

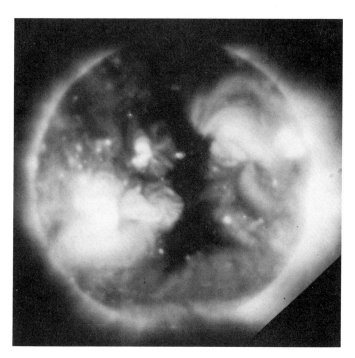

*Earth's magneto-
sphere and Van Allen
belts.*

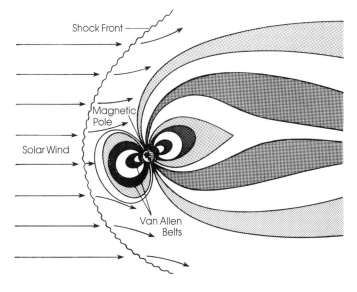

Showing how the pileup of the solar wind with the comet atmosphere produces an outer shock wave and a confused motion of the ions near the head of the comet. The result is an ion tail directed away from the Sun.

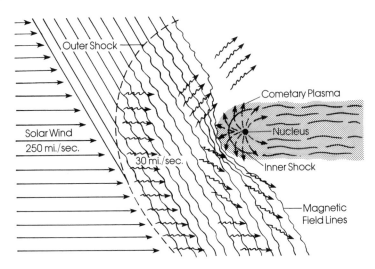

We have noted that the ion tails of comets lie in the comet's orbital plane, but curve back a few degrees from an extension of the line to the Sun. C. Hoffmeister of Germany measured this effect in 1943. Michael J. S. Belton at the Kitt Peak National Observatory and John C. Brandt at the Goddard Space Flight Center of NASA collected all available observations of comet tails to check out Biermann's theory. They found in 1969 that comet tails point like "wind socks" at airports and telltails on sailboats. The ion tails tilt with the velocity of the comet in its orbit as the comet moves across the solar wind. Cometary beards are indeed blowing in this wind, which has a velocity of 400 kilometers per second.

Acting as more than weather vanes, comet tails can provide a tool for measuring the solar "weather." As the Sun turns during its twenty-five-day rotation period, sectors of alternate magnetic polarities turn with it. The system revolves sedately like a huge expanding phonograph record. These large magnetic sectors change slowly in time, over months, whereas solar flares are small plasma storms shot out into space at irregular intervals. These changes in the solar "weather" may start or cut off the wind structure on a comet's ion tail. As a consequence, the surprising breaks in the tails of comets Morehouse (1908) and Halley really reflect drastic and sudden

John C. Brandt.

Typical alignment of magnetic fields (+, –) in the solar wind and dates that the sector boundaries were seen to cross the earth.

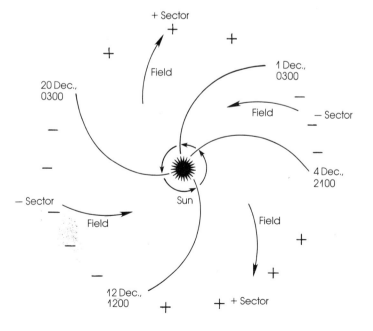

Showing progress of sector boundary: a in solar wind; b, producing a disconnection in the comet's tail because of the change in sign of the magnetic field of the solar wind.

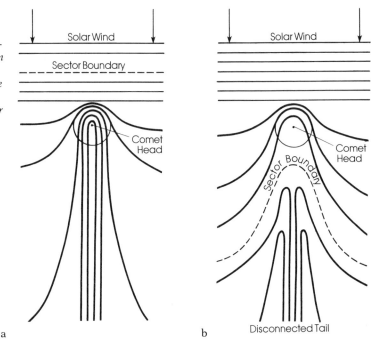

a   b

changes in the solar wind. The motions and variations in brightness of the kinks and knots reflect magnetic eddies as the solar wind sweeps by. Old records of terrestrial magnetism now make it possible to associate some of the tail activities of comets Morehouse and Halley with solar weather. The study of solar and interplanetary weather is much like meteorology, further complicated by Alfvén's magnetic fields.

On March 18, 1973, Lubos Kohoutek of the Hamburg Observatory, in West Germany, was looking for Biela's long-lost, split comet. He failed, as has everyone else, but he found a new comet at the remarkably large distance of 4.75 AU from the sun. If, as it came nearer, it had brightened intrinsically at the rate that most comets do, and some comet buffs expected it to, it would have been spectacular at perihelion, only 0.14 AU from the Sun. Because of the advance publicity it received, comet Kohoutek of 1973–74 was a huge "failure" for the public, but a huge success for cometary science. Its discovery at such a great

Disconnections in tail of Halley's comet in 1910: a, June 6, $13^h 18^m$ at Cordoba; b, June 6, $17^h 15^m$ at Mt. Hamilton; c, June 7, $10^h 46^m$ at Cordoba.

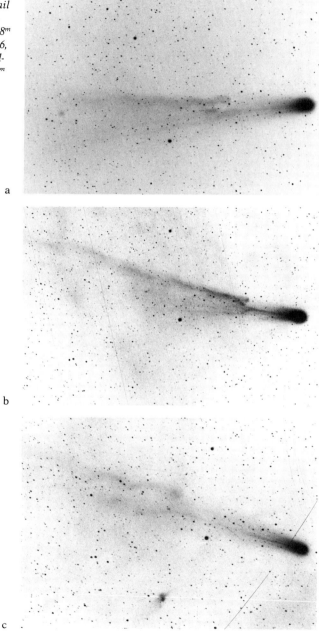

a

b

c

distance, plus the possibility of its becoming so bright, gave observers both the opportunity for unprecedented preparations and the incentive to put forth a huge effort. NASA supported scientists in planning the most diverse and comprehensive observing program ever applied to cometary study. Observations from telescopes on the ground, from satellites, from balloons, and from space probes were all coordinated. Later we shall discuss the many new results obtained from the study of comet Kohoutek.

Relevant here is the fact that its tail showed some amazing and changing structures, including a large "bend" some 97 million kilometers (60 million miles) from its head on January 20, 1974. Earth observers noted a magnetic field reversal of the solar wind during January 24–25. When they theoretically turned the flat disk of solar wind polarities backward, they found that comet Kohoutek, at 0.8 AU from the Sun, had been properly placed to meet this reversal four days earlier. All the calculations agreed, to within the expected accuracy. The geometry of the bend in the tail clearly showed just how this solar reversal had distorted the tail of comet Kohoutek. This example of "space meteorology" illustrates the success of combined ventures in space technology, physics, geophysics, and astronomy, while also demonstrating the practical complexity of cometary science.

Comets can now be used as a space laboratory for basic research on Alfvén's difficult plasma physics. Effort must be mustered to send space probes to a comet, along with simultaneous solar-wind probes, solar satellites, or shuttles, and the many ground-based observations needed to complete the program. The rewards will be a step forward in the study of the solar wind, the Earth's magnetosphere, comets, and astronomy in general. Such programs will bring more insight into radio stars, black holes, quasars, and the nature of the universe. Aristotle's "exhalations" of the upper atmosphere will truly have become exalted!

Chapter 14   **Flying Sandbanks**

The spectacular tails and great comas of comets are easily seen, but the *real* comet—the nucleus—is invisible, buried at the center of its head. When the comet is far from the Sun, we see nothing but a hazy or stellar speck of reflected sunlight. It is like a bud that blossoms forth into a flower in full sunlight. What is the hidden source of so much beauty? The theory that the nucleus of a comet is a flying sandbank held sway throughout the first half of the century. The strongest proponent became R. A. Lyttleton of England, when he and Fred Hoyle envisaged a method of making flying sandbanks from interstellar dust and gas. Their ingenious suggestion, the accretion theory, supposes correctly that the sun occasionally moves through a cloud of interstellar matter as the Solar System makes its way around our galaxy in its stately 250-million-year period of revolution.

The principle of the accretion theory is easily understood. The Sun's gravity attracts the dust of the cloud as the Sun moves through it. As the dust is drawn toward the Sun, it converges along the line following the Sun in the path of the relative motion. Because the dust is embedded in gas, it is nearly at rest in the huge interstellar cloud. Hence it converges symmetrically with the gas around the line of motion behind the Sun. Here comets are made, according to Lyttleton's accretion theory, as grains from all directions around the line of motion collide to form coherent bodies of interstellar grains with embedded gases. The velocity and the distance from the Sun will vary because the Sun passes through many such clouds over its long lifetime. This adds variety to the orbits of the comets and maintains the supply.

In support of the flying sandbank model, Lyttleton seized upon a peculiar property of the comas of some comets, the classic example being that of Encke's comet. The coma of P/Encke first grows and then tends to shrink as the comet comes closer to the Sun and as the comet brightens. For example, at 1.4 AU from the Sun, the coma

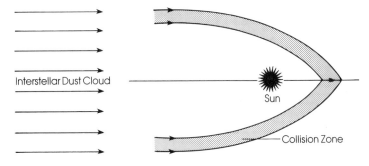

The converging inter-stellar dust and gas meet at the axis of motion all around the axis to form conden-sations.

Interstellar Dust Cloud

Sun

Collision Zone

is roughly 500,000 kilometers in diameter. Near peri-helion it may shrink to less than 16,000 kilometers. The shrinkage is by no means the same at each return, and the relation does not hold for all comets. But Lyttleton argued that this shrinkage is caused by the fact that each grain in the sandbank is moving in its own orbit about the Sun. In total, the grains are more closely packed near perihelion, where the orbits tend to converge and cross.

Lyttleton then accounted for both the gas and the long lives of comets as follows:

In the accretion theory, the dust particles of the comet must be regarded as spread right through the observable coma, and, per-haps invisibly, even beyond. This means that in the *free* motion around the Sun of all the multitudinous particles, each particle above the general orbit will cross down through it during the perihelion side of the orbit, and those below the plane will cross upwards through it. The relative speeds involved in these cross-ings can range from about 10^4 to 10^5 cm/sec [330 to 3300 ft/sec]; only a small proportion of the particles will undergo collisions as a result of this turning inside out, as it were, of the comet. Furthermore, not only are such speeds large enough to shatter colliding particles, but the collisions must produce intense local heating of the material at the tiny areas of contact. The colli-sions bringing about gas release will naturally tend to occur with greatest frequency in the neighborhood of the central regions of the comet. This would mean that somewhere within the comet there would at any one time be a centre or perhaps more than one, from which such released gases would flow out-wards . . . and possibly it is this centre and the gas flowing out-wards from it that gives the appearance of a nucleus.

One problem with this model lies in the property of sun-light to contract the orbits of small particles as they revolve about the Sun. For very fine dust, as we have seen,

Comet P/Encke. (Courtesy Yerkes Observatory.)

sunlight exerts a pressure and enlarges the orbit. For larger particles, the effective mass of light (remember Einstein) scattered by the dust also acts as a brake or resisting medium. The effect varies inversely as the diameter of the particle and is known as the *Poynting-Robertson Effect.* Hence, in a few revolutions of Encke's comet, the smaller particles would move in significantly smaller orbits with shorter periods than the larger particles. The former would forge ahead of the larger particles to stretch out the nucleus along the orbit. Also, as E. Schatzman demonstrated in 1952, the sandbank nucleus would, over long periods of time, coalesce into a discrete mass because of the continual collisions among the particles held together by their mutual gravity, but otherwise moving freely in deep space.

The sandbank model is quite hopeless for the sungrazing comets. In the 1920s, Henry Norris Russell, the great astrophysicist at Princeton University, noted that rocks 1 foot in diameter would be boiled away in the few hours during which a sungrazer passes through the inner corona. In response, Lyttleton wrote, as late as 1962,

> But even if the comet were completely vaporized at perihelion (the sungrazing comet 1882 II elongated into a brilliant streak and shortly after perihelion divided into five or more pieces following each other along much the same orbit—it became known as the "string of pearls" comet—which does not suggest the motion of large blocks in a nucleus a few kilometers in diameter) there seems no reason why, on receding from the Sun again, it should not recondense into solid dust particles.

We know today that both the solar radiation and the solar wind dissipate any gas near the Sun, preventing it from ever condensing as it is blown away into deep space.

The persistence of comets creates a problem for the sandbank model, in spite of Lyttleton's optimistic appraisal. The rate of release of gas from grains should surely fall off rapidly, leaving the comet a cluster of particles, reflecting only the Sun's spectrum. But Encke's comet, which has certainly made hundreds of revolutions, shows gases in its spectrum almost entirely, with only a trace of directly reflected sunlight. Could a comet replenish its supply of gas when far away from the Sun? The Sun actually does move through interstellar clouds of dust and gas. Most of these clouds, however, are almost a pure vacuum, better than the best laboratory vacuums.

Diameter of comet image depending upon photographic exposure, from 1 to 8 seconds. Comet Kohoutek, January 15, 1974. (Courtesy Lowell Observatory.)

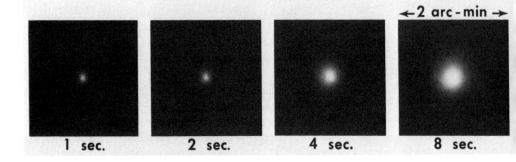

←2 arc-min→

1 sec.   2 sec.   4 sec.   8 sec.

Typically, they may contain one to a few hydrogen atoms per cubic centimeter, and only a small percentage of that number will consist of heavier atoms. This raises the question of how much gas a comet actually loses on its way around the Sun, and therefore, how much gas it would have to capture to remain healthy.

This question remained unanswerable until physicists began to understand how atoms absorb and radiate light. In 1911, the German astronomers Karl Schwarzschild and E. Kron first attacked the problem of gas production in comets. Their results, revised later by Karl Wurm in 1943, gave Halley's comet in 1910 a mass production rate for C_2 of about a fifth of a ton per second. This much C_2 was required to provide the observed brightness of the bands before the sunlight split the C_2 molecules. Only a few kilograms per second of the carbon monoxide ion was needed for the great tail. The large amount of gas lost by comets thus became an additional problem for the accretion model. Sandbank comets simply ran out of gas, and had no known replenishing supply.

Another defect of the sandbank model has to do with the peculiar property of P/Encke's deviation from Newtonian motion, that is, its decreasing period at successive revolutions. A resisting medium might conceivably explain P/Encke's motion, but then could not also explain the fact that some other comets have increasing periods. Furthermore, by the 1940s there was reliable evidence that the amount of gas or dust near the Solar System was too small to affect the motions of comets.

By midcentury it was obvious that the flying sandbank model of a comet had become inadequate. The model would quickly lose any residual gases it might have retained and had no source in space to replenish them. It could not survive as a sungrazer. It could not account for changes in the period of revolution about the Sun. Another theory of the cometary nucleus was clearly needed.

Chapter 15   Dirty Snowballs

When I returned to research after the three-year distraction of World War II, my continuing studies of photographic meteors brought me face to face with comets again. The double-station triangulation of meteors gave their velocities and orbits. Meteor orbits were distinctly cometary, not like those of the numerous asteroids inhabiting the space inside Jupiter's orbit. Almost all of the meteors appeared to come from comets, and there was no evidence of interstellar interlopers among them. Furthermore, comet Encke appeared to be the parent of a huge volume of meteoroidal bodies in related orbits. Comets seemed to provide most, if not all, of the small particles inhabiting our local interplanetary space. These meteoroidal particles have rather short lives because of collisions and their spiraling toward the Sun by the Pointing-Robertson effect, lives measured in tens of thousands of years, not even millions. I calculated that only a few tons per second of cometary debris could maintain this complex of interplanetary particles. Between lecturing and meteor research, I repeatedly wrestled with the enigma of comets, their nature, and their origin, along with that of the major bodies of the Solar System.

'In the late 1940s, only a handful of astronomers in the world were giving much attention to the local problems of the Solar System. Gerard P. Kuiper, an eminent astronomer of Yerkes Observatory, placed the U.S. contingent at 1.5 full-time astronomers in planetary research, an ironic underestimate, but still symptomatic. I decided to leave the Universe to other astronomers and concentrate on problems in my own backyard.

The clues about the nature of comets were numerous. The flying sandbank model failed to unlock the mystery because it could not explain how comets could wantonly waste such a large amount of gas into space without having a continuing supply or a source for replenishment. Furthermore, in the 1940s, not only CN and C_2, but also the radicals OH, C_3, and NH had been identified in the

146

heads of comets, while the positive ions CO^+, CH^+, OH^+, N_2^+, and NH^+ could be found in the tail. Many heavier atoms were clearly present in the dust, and observed as meteors. Pol Swings of Belgium and other astronomers noted that OH might come from water (H_2O), CH from methane (CH_4), NH from ammonia (NH_3), and CO from carbon dioxide (CO_2). All of these are stable molecules that can be "put in bottles" and are made of elements that are relatively abundant in the Sun. Water, methane, ammonia, and carbon dioxide seemed to be likely *parent* molecules. Because they do not show strong spectra, we could see only their broken pieces, even if they were present in comets.

It became obvious to me in the late 1940s that comets must carry a large reservoir of these parent molecules to keep some comets active for hundreds, or even possibly thousands, of revolutions about the Sun. In addition, some comets must be big enough and solid enough to graze the Sun without experiencing total destruction. The answer was clear: The nucleus of a comet must be a great mass of ices embedded with dust or meteoric particles— in other words, it must be a huge, dirty snowball. Such a snowball would remain dormant at great distances from the Sun, becoming active only when near enough for sunlight to vaporize the ices. Actually, in a vacuum, ices do not melt to form a liquid. They simply evaporate at the surface; technically they *sublimate*. The sublimation of snow is quite obvious in cold climates on Earth, where a thin layer will disappear in a day or two at temperatures well below freezing. On a comet, the sublimating ices would lose molecules to space at a speed of more than 300 meters per second, according to the kinetic theory of gases. Their pressure would be quite adequate to carry away any embedded rocky grains. Thus the dirty snowball could produce both the gas and the grains observed to come from comets.

The idea that comets might be made of ice goes back to the great natural philosopher and mathematician Simon Laplace, who lived in the seventeenth century. He suggested that the Sun might vaporize ice when comets come close, and that the vapor might freeze again as comets recede to the bitter cold of deep space. His suggestion was

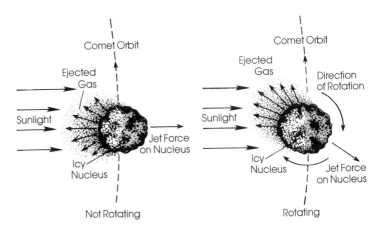

Jet action on a cometary nucleus. Left, if the nucleus is not rotating, the jet force is directed in a direction opposite to the Sun; right, if the nucleus is rotating, the jet force may tend to retard the motion around the Sun (as indicated in the diagram.)

ignored when comets were shown to produce meteor streams, and when physicists found that cold solids can hold a bit of gas, which can be released when heated. The activity of the prodigal Sun, of course, blows away the gases in space, preventing any significant amount of vapor from freezing again on the comet.

Having concluded that comets must be dirty snowballs, I felt that the idea was both obvious and difficult to prove—just another theory about comets that merited only a passing comment in some popular article. But I kept recalling Encke's comet and its erratic motion, which defies Newton's law of gravitation. What could change its period? At that time, in 1949, I was attempting to improve the theory of atmospheric resistance to a meteor plunging through the atmosphere at tens of kilometers per second. The front of the body is boiling away because of the air friction. I attempted to formulate the additional drag on the body introduced by the jet action of the gas from this boiling surface. Suddenly I realized that the gases sublimating from a dirty snowball act in identical fashion. A snowball in sunlight is really a small jet engine. As the ice sublimates on the sunny side and the

ASTRUM HAROLDO PRAEFIGURATUM A.D.1066 = PRO STELLAE RESURRECTO A.D.1986 DELPHINE DELSEMME FEC

Bayeux tapestry depicting Halley's comet in 1066. (Duplicated by Delphine Delsemme.)

Adoration of the Magi, *painted around 1303 by Giotto di Bondone; located in the Arena Chapel in Padua. The comet depicted as the Star of Bethlehem was very likely Giotto's impression of P/Halley, which appeared in 1301. (Courtesy Donald K. Yeomans.)*

Left: *Donati's comet over Paris, October 4, 1858. (Amédée Guillemin,* Le Ciel *[Paris, 1877].)* Lower left: *Cheseaux's comet over Lausanne, March 8, 1744. (Guillemin,* The World of Comets *[London, 1877].)* Lower right: *Sungrazing comet over Paris, March 19, 1843. (Guillemin,* The World of Comets.)

Comet Ikeya-Seki in the morning sky of October 31, 1965, photographed from 7,000 feet in the Catalina Mountains near Tucson, Arizona, with 50-millimeter f/1.9 Miranda camera; about a 20-second exposure on high-speed Ektachrome. (Photograph by Dennis Milon.)

Comet Bennet 1970 II, showing hoods in the inner coma. Composite of four images. (Obtained by S. L. Larson, Lunar and Planetary Laboratory.)

*"Halley's Comet Rag,"
by Harry J. Lincoln.
Sheet music cover,
1910. (Courtesy Ruth
Freitag, Library of
Congress.)*

*People's reactions to
comets have ranged
from illness and ter-
ror to inspiration, as
these press clippings
show.*

Some panicked, some profited as the fiery comet closed in

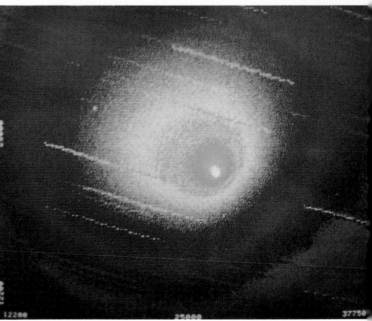

Computer enhanced photograph of Comet Halley as seen in 1910. Photo courtesy of the National Optical Astronomy Observatory/Lowell Observatory.

Comet IRAS-Araki-Alcock 1983d, false-color computer image obtained from a 20-minute exposure with a 90-centimeter Schmidt telescope at Asiago Observatory on May 9. (Courtesy C. Cosmovici.)

The electromagnetic spectrum showing the range of visible light. The wavelength is shown on the logarithmic scale. Courtesy P. A. Jacobberger, National Air and Space Museum.)

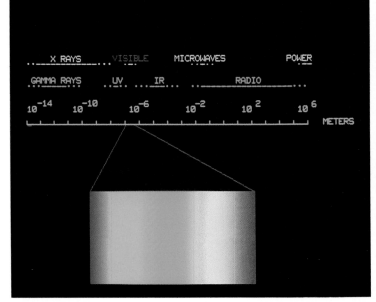

Infrared image of the center of our Milky Way Galaxy, produced from observations by the Infrared Astronomical Satellite (IRAS). The knots and blobs scattered along the bank (the galaxy's disk) are giant clouds of interstellar gas and dust. Many are stellar incubators. (Courtesy Jet Propulsion Laboratory.)

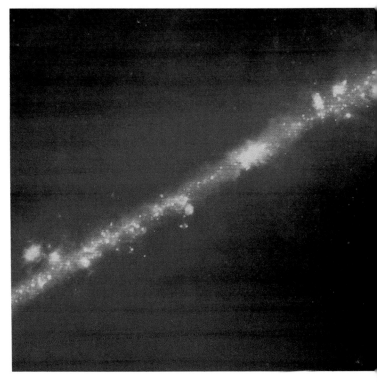

The great daylight fireball of August 10, 1972, as it appeared over Jackson Lake in the Rocky Mountains of Wyoming. This tiny asteroid skimmed the Earth's atmosphere before continuing on into space. Such large objects occasionally do hit Earth. (Photograph by James M. Baker; courtesy Dennis Milon.)

False-color image of the star-formation region in the constellation Orion, obtained from IRAS observations. (Courtesy Jet Propulsion Laboratory.)

The Skylab space station as photographed in 1974 by the final crew before returning to Earth. Observations and sketches of Comet Kohoutek were made by the astronauts aboard Skylab 4. Photo courtesy NASA.

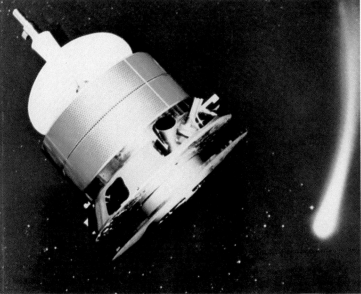

The Giotto spacecraft will return the first images of the nucleus of a comet when it encounters comet Halley in March, 1986. It will fly within 500 kilometers (310 miles) of the nucleus, and will return data on cometary dust and solar wind interactions. (Photo courtesy European Space Agency.)

molecules leave the snowball at high speed into the vacuum of space, each molecule kicks back on the surface of the snowball. This is the basic principle of jet propulsion, or Newton's principle of action and reaction.

If the comet snowball were not rotating, it would be pushed directly away from the Sun by the jet effect, which is a very slight force compared to the Sun's attraction and is almost impossible to detect from orbit calculations. But, all astronomical bodies appear to rotate, one way or another! A point on a rotating snowball will have its daily cycle of night, sunrise, high noon, and sunset. At night, in the vacuum of space, the temperature will drop precipitously, and warming will be slow in the morning. The temperature rises to a maximum sometime after high noon, as it does, in fact, on Earth. The reversed jet action, averaged over the day, will not point directly away from the Sun, but will be tilted by some angle to the line from comet to Sun. If the comet is rotating in the same sense as it moves around the Sun (that is, prograde), the jet force will push it forward in its orbit and systematically increase its period of revolution about the Sun. For Encke's comet, the orbital period is observed to decrease. Therefore, Encke's comet must be spinning retrograde compared with its orbital motion.

When I put some numbers into a simple form of this theory, I found to my delight that if Encke's comet were only a few kilometers in diameter, and rotating something like the Earth (but in the opposite sense), with an hour or so delay in the heating, then the sunlight could vaporize enough ice to provide the kick required to decrease its period of revolution, as observed. The comet need lose less than 1 percent of its mass per revolution, perhaps a meter or so of its radius. It could survive at least hundreds, if not thousands, of revolutions about the Sun.

Happily for the theory, other periodic comets, P/d'Arrest and P/Brooks 1, as well as possibly P/Halley, were also observed to deviate from proper Newtonian motion. Their periods were increasing, in contrast to the period of comet Encke. They must be spinning in the sense of their solar revolution. As I was preparing my paper for publication, I found that, in the USSR, A. D. Dubiago had calculated the jet action required for these and three other comets, but like Bessel, more than a century before, he did not prescribe a specific physical mechanism. He considered that the gas jet action was inadequate and that the explosive expulsion of solid particles must produce the effect. It is interesting to note that, of these first six comets, three show period decreases and three increases. The spin of the comet snowballs seemed random, half prograde and half retrograde. This randomness has prevailed among more than three dozen such comets studied since then by Brian Marsden and his colleagues at the Smithsonian Astrophysical Observatory.

I began to wonder how powerful my snowball model might be in blowing off solid pieces to make meteor streams. The force of gravity on an icy ball a few kilometers in diameter is extremely small. A particle given a speed of about a meter per second, for each kilometer diameter of the snowball, will sail away into the depths of space, never to fall back. With my simple theory, I calculated that, if P/Encke is 5 kilometers in diameter, it could—at perihelion—blow away rocky pieces as large as 3 meters in diameter, which would be quite adequate to make meteor streams.

The snowball comet also should withstand the intense heating that is experienced by the sungrazing comets. The loss of ices might amount to many meters of radius in such a passage, yet sufficiently large snowballs could theoretically survive. But if they were quite large and weak structurally, another physical effect could occur: The snowball might be torn apart by the Sun's huge gravity field. It might split, precisely as some of the sungrazing comets have been observed to do. The process of "tidal" splitting comes about because the Sun's gravity is

The split nucleus of the sungrazing comet of 1882. Left, October 13; right, October 17. (By Edward Holden.)

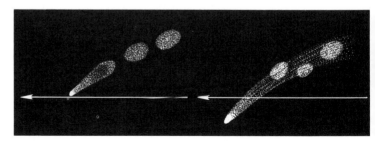

stronger on the Sun side of the sungrazing body than on the far (dark) side, falling off according to Newton's inverse-square law of the distance. Thus the tidal force tries to pull the comet apart. The larger the snowball and the weaker its internal strength, the more easily it can be torn apart. Both Öpik and I have calculated independently that a huge snowball sungrazer, some 60 kilometers in diameter, could be split by the Sun's gravity if it were considerably weaker than ordinary ice, perhaps something like a frozen sandbank. Such a comet may have been the original ancestor of the sungrazing group of comets. Smaller descendants of the sungrazing patriarch could well survive such passages if not too cracked or weakened internally. The loss of small pieces, as observed, fits well with the theoretical expectations.

The dirty snowball model was accepted rather quickly by most of the few astronomers interested (with one notable exception, mentioned in the previous chapter). The question of the origin of comets aroused immediate interest. When and how in the universe can one make dirty snowballs? Oort, in his discussion of the Öpik-Oort reservoir, or comet cloud surrounding the Sun, suggested that comets may have originated with the asteroids in the inner Solar System, and subsequently were kicked out into the great cloud by gravitational disturbances from the giant planets. Most astronomers, however, favored the idea that the inner Solar System had always been too hot for ice to freeze, particularly within Jupiter's orbit, where the rocky asteroids abound. Much lower temperatures were required if ammonia and carbon dioxide were also to freeze. Hence it seemed reasonable to theorize that comets were formed in the outer reaches of the Solar System, along with, perhaps, Uranus and Neptune. They could then be kicked out more easily, still in a very wasteful fashion, to the far reaches of the Öpik-Oort cloud. More recent thinking about these evolutionary problems will be discussed in a later chapter. Here it suffices to say that in the 1950s, the dirty snowball model seemed consistent with the assumption that comets originate somewhere in or beyond the outermost planetary regions.

A problem that arises in postulating ices in comets is the possibility that the extremely volatile methane (CH_4)

may be a parent molecule for CH. Possibly, ammonia and carbon dioxide could be frozen out at rather low reasonable temperatures, but, even at 20 degrees above absolute zero ($-459°F$), methane remains a problem. The Belgian comet astronomers Armand Delsemme and Pol Swings introduced the idea that the methane was perhaps not frozen at all, but that its molecules were simply trapped among the water-ice molecules. To the chemists, such a mixture in the laboratory is known as a *clathrate*. As we shall see, the details of the chemistry of the comet ices still constitute a major problem, perhaps one that can be solved only by space probes to comets.

The dirty snowball concept proved useful in the theory of cometary phenomena by helping to clarify the physics of Bessel's fountain theory and of Bredichin's Type II (III) comet tails, the dust tails. In 1957 comet Arend-Roland blossomed forth with one of the most beautiful dust tails to be seen since the unexpected comet 1910 I upstaged the return of Halley's comet. By 1957 the theory of light pressure on fine dust had been well developed, and the tricky problems of dust propelled by gas in sonic and supersonic flows were also well understood. This enabled two experts in the related field of aerodynamics, Michael L. Finson and Ronald F. Probstein of the Massachusetts Institute of Technology, to interpret some photographs of comet Arend-Roland made by Zdenek Ceplecha of Czechoslovakia. They analyzed these photographs to find out just how such comet tails develop and the size of the dust particles involved. Actually, they perfected the fountain model of comets, which was introduced by Bessel and Bredichin, one of our clues about comets. Finson and Probstein found that, at maximum activity near perihelion (0.32 AU from the Sun), comet Arend-Roland was ejecting some 10 tons of gas per second, which was carrying along with it an even greater mass of fine dust. From the rate at which solar light pressure repulsed the dust, they deduced that the dust was extremely fine, most particles having diameters about one-quarter the wavelength of red light and tailing off to relatively few particles as large as 0.1 millimeters (1/250 inches).

Such a mass loss to the comet adds up to about 2 million tons per day, and, to about a hundred times this much during the total apparition. If we spread this seemingly huge loss over a snowball that is, say, 6 kilometers in diameter, the comet loses a layer about 1 meter deep, as is consistent with the amount of ice that sunlight theoretically can sublimate. Comet Arend-Roland would scarcely notice this loss and would be able to return many times before it began to weaken. For other reasons, however, comet Arend-Roland will never return. It came toward the Sun in an almost parabolic orbit. The major planets have already kicked it into a hyperbolic orbit, so that it will be lost forever in the depths of our Galaxy. Alas!

Comet Arend-Roland startled comet watchers by being the brightest comet for many decades to show a conspicuous tail pointing toward the Sun. This sunward tail became an extremely thin spike, pointed precisely at the Sun and extending for some 18 degrees, at the moment when the Earth crossed the comet's highly tilted orbital plane. No mystery here: Larger hunks of meteoric-type material had been ejected with low speeds and were moving almost exactly in the orbital plane. When the Earth moved into this plane, observers could see the sunlight reflected from huge numbers of these bodies stretched out along the orbit. This would have made a fine meteor shower if the Earth had actually crossed the path of the comet.

In 1968 Biermann, predictor of the solar wind, made another important prediction on the basis of the dirty snowball model. Together with a colleague, he had made calculations about comet Arend-Roland, much along the lines described by Finson and Probstein. He believed that the large amount of gas produced by comets must come mainly from water-ice. The OH bands are in the ultraviolet and are hard to observe, but they indicate a rather large amount of the parent molecule H_2O. The comet would be losing water vapor at a rate of several tens of tons per second near perihelion. Ultraviolet sunlight rapidly breaks up water molecules into OH, oxygen, and hydrogen once the water leaves the comet as a gas. In the near-vacuum of space, the hydrogen remains atomic and neutral for about three weeks at 1 AU from the Sun. Its fluorescent spectral line, called *Lyman Alpha*, lies in the extreme ultraviolet region of the spectrum, where the

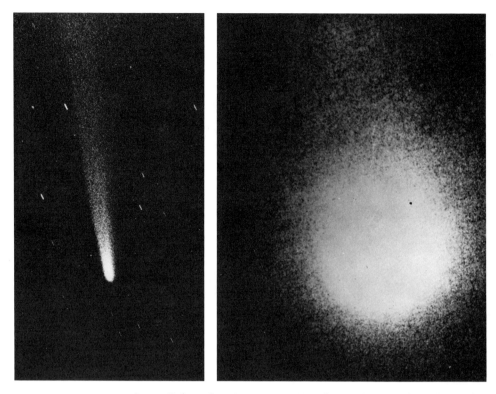

Comet Kohoutek on January 8, 1974. Left, in ordinary light; right, on the same scale in far-ultraviolet hydrogen light. The hydrogen image is larger than the Sun. (From Aerobee Rockets by C. B. Opal, G. R. Carruthers, D. K. Prinz, and R. R. Meier.)

Earth's atmosphere is as opaque as a piece of lead. Only spacecraft above the Earth's atmosphere could possibly spot it.

On January 14, 1970, when the 2nd Orbiting Astronomical Observatory turned its ultraviolet detector in the direction of comet Tago-Sato-Kosaka, the comet was a little more than ½ AU from the Sun. Arthur D. Code of the University of Wisconsin detected a huge cloud of hydrogen atoms around the comet. The spectacular comet Bennett presented an almost unbelievable cloud to the 5th Orbiting Geophysical Observatory in 1970: The hydrogen cloud could be traced over a diameter greater than 3 million kilometers, or about three times the diameter of the

Sun! The cloud was expanding at a rate of more than 8 kilometers per second.

The fast expansion of these hydrogen clouds is well understood. In the late 1950s, L. Haser of Belgium developed the basic theory explaining how sunlight is able to crack up molecules around comets. In the case of water, as we have noted, the ultraviolet light tears the H_2O molecule apart into OH, H, and O. When the OH radical is later broken apart, it throws out the hydrogen at about 8 kilometers per second, thereby largely controlling the expansion speed of the hydrogen clouds. Because the oxygen atom weighs sixteen times more than the hydrogen atom, it is kicked back at only 1/16th the speed of the hydrogen in the more slowly expanding coma. The theoretical and measured temperatures of the clouds are some 1,100°C.

Even "tame" comet Encke shows a hydrogen cloud, but it is not nearly as large as those from the bright, "wild" comets, which include the infamous comet Kohoutek of 1973–74. At the standard distance of 1 AU from the Sun, P/Encke loses only about 0.1 ton of water per second, as measured by the H and OH, whereas the great comet Bennett of 1969–70 and the split comet West of 1976 lost about 14 tons per second. Halley's comet probably lies near this upper bracket.

Finally, we must clarify the question of why the comas of comets shrink as they approach the Sun, a phenomenon very strongly stressed by Lyttleton in defense of the sandbank model. Actually, the comas are almost always very small at great distances from the Sun. They tend to reach their maximum diameters at 1.5 to 2.0 AU, and usually tend to shrink as they approach perihelion. The original expansion, as comets come in from great distances, occurs when the Sun begins to vaporize the ices. We see the gases coming out of the coma until the molecules are broken apart by ultraviolet sunlight. Hence the apparent diameter depends on how far the molecules can go before they become invisible. This distance is simply the product of their lifetimes times their speeds of expansion. The speeds increase slowly with decreasing solar distance, and the lifetimes decrease as the inverse square of this distance because the destructive sunlight increases in the same manner. Thus, once the comet becomes active on approaching the Sun, the lifetimes of the fluorescing molecules decrease faster than their speeds increase, and the coma shrinks. The actual shrinkage observed is highly erratic from comet to comet, and cannot be generalized into this simple law. In principle, however, there is no longer any mystery about coma diameters and how they vary with solar distance, but a number of intriguing problems remain to be solved.

Our comet clues have finally led us to the assumption that a dirty snowball is the nucleus and basic component of a comet. This model clarifies many comet mysteries, such as: the longevity for large comets but short lives for small ones; both the survival of some sungrazers and the splitting of others; orbital motions that deviate from Newton's law of gravity; the origin of cometary meteor streams; the origin of the coma fountains—which are basic to the Bessel-Bredichin theory of comet tails; and the general nature of cometary spectra, both fluorescence of gases and reflection from dust. The model does not specify the place of origin. The nature of the snowball, however, requires that comets be formed in very cold and relatively dense regions of space, a requirement that fits in with our ideas of the outermost regions of the Solar System when it was coelescing, or even of regions in very dense and cold interstellar clouds.

The place and time of origin and a great number of other problems about comets will occupy the attention of astronomers for many years to come. For the most part, these problems require complicated analysis involving both observation and theory.

Chapter 16 **Spinning Comets**

The idea that comets are really cosmic snowballs of ice and dust became a springboard that enabled me to learn about many normal comet activities and some idiosyncratic behaviors. Especially intriguing to me were the beautiful "parabolic" halos shown by several comets, particularly Donati's of 1858, described earlier. Why, I wondered, do the halos seem to recur with almost clockwork precision? A spinning nucleus, markedly different on each of its two faces, could explain the observations. Suppose part of one side or hemisphere is covered with ices while the rest of the nucleus, for some unknown reason, is covered with thick dirt, or meteoric stuff. During each "comet day," the icy area would throw out its vapor and grains, while the rest of the nucleus would do nothing at all, there being no ice at the surface to produce such activity.

At first I had assumed that comet nuclei are uniformly round. But why should they be perfectly round snowballs, and exactly the same all over their surfaces? (Shades of Aristotle?) In all likelihood, they originally grew by accretion, from the impacts of smaller snowballs, beginning with dust and cosmic snowflakes of some sort. Is there any reason to think that all of them should be alike? Of course not! Some might have been formed in warmer regions, and might have been composed predominantly of rocky grains. Possibly the grains grew first and then the temperature fell, so that they became covered with snow or frost. Or perhaps all the particles were originally interstellar grains that contained ices. The purpose of studying comets is to help reconstruct these primitive conditions, which were probably prevalent at the time when our Sun, Earth, and entire Solar System were being assembled. Perhaps comets can tell us something about those ancient times, when the atoms now in our bodies were floating around in a cosmic cloud.

Each evening in early October 1858, George P. Bond at Harvard looked through the Great Refractor and

157

Coggia's comet of 1874 on July 13. (Drawn by Brodie.)

Tebbutt's great comet of 1861. (Drawing by J. F. Julius Schmidt, Vienna.)

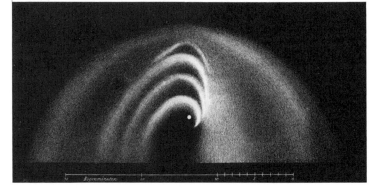

observed the halos of comet Donati, which became slightly larger from one evening to the next. He concluded that they were being thrown off about every five to six days. Meanwhile, in Vienna, J. F. Julius Schmidt, a great double-star observer, carefully measured the diameters of the halos across the coma perpendicular to the direction of the tail. He made five to ten settings of his telescopic crosswires for each measurement. From these measures, he concluded that the halos were growing quite rapidly and that a new one appeared every few hours. To explain the halo formation on the basis of the spinning snowball model, I needed to know how fast such envelopes were expanding.

I found that Nicholas T. Bobrovnikoff, the most thorough student of comets in the world, had, in 1954, already calculated the expansion speeds for the halos of some twenty-seven comets on the basis of numerous observations by many individuals. The average was roughly what I had expected for grains carried by subliming ice vapor,

Donati's comet of 1858. (Drawn by Otto Struve at Pulkovo on October 5.)

about 0.5 kilometers per second (1,800 feet per second) at 1 AU from the Sun, and falling off slowly with solar distance. From the geometry and velocity, I could calculate the time taken for each halo to reach its observed diameter. Subtracting this from the time of each observation, I had a series of instants, or *zero dates*, when each halo started out from the nucleus. If these zero dates each measured the start of activity on a spinning, lopsided, or nonsymmetrical snowball, they should be separated by multiples of the period. Indeed they were. The period for comet Donati came out to be 4 hours, 37.2 minutes. The nucleus was spinning like a clock, in fair agreement with one of Schmidt's several original period calculations.

An amazing coincidence probably explains why Bond and most other astronomers rejected Schmidt's short period for the production of the envelopes on Donati's comet. On the globe of the Earth, the west longitude of Harvard Observatory from Greenwich, England, is 4 hours 44.5 minutes, only 7 minutes greater than the comet's rotation period. When astronomers in England or western Europe looked at the comet each evening at roughly the same solar time, they had almost exactly the same view as did Bond, one period later. The envelopes were nearly identical for a couple of weeks, even to fine details. Also, Bond did not credit Schmidt with the precise measures he actually achieved because Schmidt's telescope had only a third the diameter of Bond's. The Harvard refractor was one of the two largest lenses in the world, its twin being at Pulkovo near St. Petersburg, now Leningrad.

Be that as it may, the clockwork precision of halo production from Donati's comet over more than seventy-eight

Donati's comet of 1858. Left, *October 6;* right, *October 7. (Drawn by J. F. Julius Schmidt, Vienna.)*

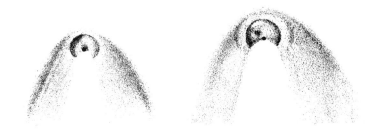

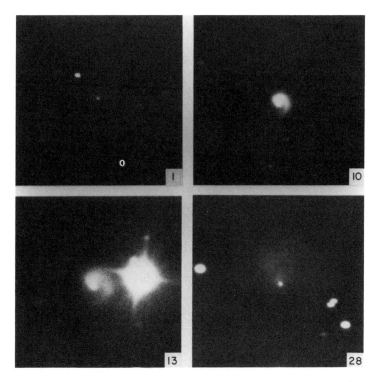

*Sequence showing an outburst of P/Schwassmann-Wachmann 1, as pho-
tographed on February 1, 10, 13, and 28, 1981, by R. E. McCrosky and C.-Y.
Shao at Oak Ridge Observatory with the 1.5-meter reflector, operated by a
grant from NASA. Compare this outburst with that of the same comet in
1961, shown in Chapter 20.*

periods clearly required a rotating, spinning comet
nucleus that was very active on one side and quite inactive on the other. Therefore the actual snowball nucleus
had to be asymmetrical over its surface.

Its fast spin led me to question whether the nucleus
might fly apart because of the centrifugal force at its
equator. Actually, the nucleus needs to be fairly round to
hold together. For a spherical ball of pure water-ice, the
centrifugal force at the equator will exceed the gravity if
the period is less than 3.3 hours, whatever the diameter. If
the dirty snowball is denser than water-ice because of the
dust, the critical period is smaller. Nevertheless, the
nucleus of Donati's comet, for example, cannot be shaped

like a brick unless it is quite strongly made, without cracks or weak seams.

Oddly enough, the first two comets for which I calculated the spin periods represented nearly the extremes in the range of periods that I later found for comets in general. The other was the phenomenal periodic comet Schwassmann-Wachmann 1—P/SW1 for short—which moves just beyond Jupiter in a nearly circular orbit. It has been a mystery since its discovery in 1927. Although a very large comet, it is usually very faint, observable only by large telescopes, and then not visually. Yet, now and then, it flares hundreds of times in brightness, becoming an object easily seen in small telescopes. I found that its spin is quite stately, almost exactly five days per turn, making it the longest comet period I have measured. The average period is about fifteen hours.

P/SW1 presents a real puzzle, and also a key, perhaps, to the nature of comets. Water-ice at 6 to 7 AU from the Sun, even in the vacuum of space, ignores solar heating, and will not sublimate. It might as well be rock. More exotic ices than water-ice are clearly needed to explain any activity at the distance of P/SW1, not to mention its mad outbursts. We will return to P/SW1 after we study other clues about the unearthly composition of comets.

While I was first calculating the spin periods of these two comets, Zdenek Sekanina, my colleague at the Smithsonian Astrophysical Observatory, independently applied the dirty-snowball model to the spin of comets. He realized that, if the gas and grains from a spinning nucleus come out mainly from the afternoon face of the comet, the asymmetry should show up both in good comet photographs and in visual observations. Indeed, this is true for many comets (as I had found for comet P/SW1). The inner coma streaks out in a direction that gives a clue to the direction of the spin axis, the two being more or less perpendicular to each other as seen from the Earth. In many cases, the geometry of Sun, Earth, and comet hides the effect. Sekanina also realized that asymmetry in the coma could measure the afternoon delay of the principal activity, or the *lag angle* of sublimation.

In 1979, Sekanina published his calculations of the direction of the spin axes for four short-period comets.

Zdenek Sekanina.

One of them was our old friend, P/Encke, the "Rosetta Stone" of comets. As usual, comet Encke was unusual. Its spin axis had the expected tilt that made its spin opposite to its motion in orbit. The surprise was that its polar axis lay only 5 degrees from the orbital plane and its projection on the plane only 1 degree from the direction of perihelion. The midnight sun on comet Encke shines down almost overhead at its south pole about three or four days after perihelion. Sekanina also showed that comet Encke delays its maximum gas production about 45 degrees from noon, or until about 3 p.m. in the comet's afternoon.

As an aside, the convention for labeling directions in bodies in the Solar System is to call the pole up, or north, if the rotation is counterclockwise, looking down onto the pole. Clocks, of course, were invented in the Northern Hemisphere on Earth. The hour hand was made to turn in the sense that the Sun appears to move during the day for an observer facing south. If our culture had developed in the Southern Hemisphere, clocks undoubtedly would turn the other way, and the South Pole would be considered up. The Earth would still turn counterclockwise.

The observed reduc-
tion in the change of
period of Encke's
comet since its discov-
ery. The curve gives
the loss in period for
each revolution of 3.3
years.

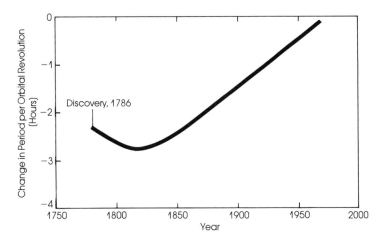

Encke's comet is peculiar in still another way. It brightens up to maximum light before perihelion, in contrast to
most comets. P/Halley, for example, is brightest somewhat after perihelion. Apparently, the south polar regions
of Encke's comet are relatively dead compared with the
equatorial regions and the northern hemisphere, because
the comet fades so fast when the southern hemisphere is
turned toward the Sun shortly after perihelion. Probably
the southern surface is covered with meteoric debris, or
perhaps the comet originally had a rocky southern hemisphere. Note that comet Encke is old, so that we probably
are looking at the inner core of a nucleus that was once
very much larger than it is today.

Once Sekanina had located the pole of Encke's comet
and found it so strangely placed (almost parallel to the
long axis of its orbit), we independently realized that this
peculiarity might help explain its major remaining mystery—the rapid change in its orbital period a century and
a half ago, which slowly decreased almost to zero by the
1970s. Suppose the nucleus were not round, but somewhat oval shaped, a flattened ball. Internal forces and
friction in that case would make it spin about its shortest
axis. It would spin with its bulge at the equator, like a
squashed top. The jet forces that change the period would
also try to tip the spinning top or nucleus, except when

the Sun's rays were shinning directly on the equator or a pole.

All spinning tops respond to a tipping force. The result is known as *precession.* Gravity makes a top fall over when it is not spinning fast enough. Once set spinning, the top keeps wobbling around, its spin axis turning around the vertical. The Earth is a familiar astronomical example of a disturbed *gyroscope,* or spinning top. In the second century B.C., the great Greek astronomer Hipparchus had set out to make a new star catalogue. He discovered, to his surprise, that the positions of the stars had shifted appreciably in the 150 years since an earlier Greek catalogue. The Earth's polar axis was shifting on the sky! Our current north pole star, Polaris, was then some 15 degrees away from the true pole. Actually, the Earth's pole appears to swing around among the stars on a circle with a radius of 23½ degrees, completing the circuit in about 26,000 years—the East Indian Year of the Gods, for which each day is a man's lifetime of 70 years. This motion is known to astronomers as the *precession of the equinoxes.* Many centuries after Hipparchus, Sir Isaac Newton was able to explain the Earth's polar motion, or precession, when pendulum gravity measures showed that Earth is flattened at the poles. The Moon and Sun attract the equatorial bulge of the spinning Earth and try to turn it

For the oblate comet nucleus in the diagram, the jet force is directed below the center of gravity of the nucleus and so tends to tip the pole counterclockwise, causing precession.

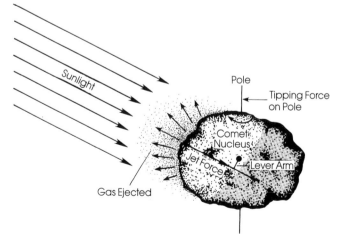

toward the plane of the Earth's orbit. Because the Earth is spinning like a top, the pole turns or precesses perpendicular to the force and keeps its angle with the Earth's orbit almost constant, making the tilt of the Earth's equator to its orbit 23½ degrees.

Sekanina and I joined forces to see whether we could apply these ideas to comet Encke and make sense out of its strange motions. We coded our computers to calculate the jet forces and the motion of the pole for an oval nucleus that is relatively dark and inactive over its southern hemisphere. We found that we needed to know the spin period, which I found to be about 6½ hours. To our great satisfaction, our calculations fit very well with the observed variable changes in the period. The nucleus came out slightly flattened, a small percentage of its radius at the poles. The calculated amount depended upon the diameter we assumed, which we took to be about 3 kilometers. If the actual diameter is greater than this, the flattening will increase proportionately.

Most exciting was the large motion of the pole across the sky, almost a degree per year over the nearly two centuries since Encke's comet was discovered. To our surprise, the pole apparently had become almost stuck in one direction for several hundreds of years before 1700. During this early time, the south pole was turned away from the Sun when the comet was most active near perihelion. We speculate that this odd circumstance may explain why the southern hemisphere of the nucleus is dead. Most of the activity took place in the northern hemisphere for hundreds of years. Solid pieces of the comet, too large to

The wobble of precession. Left, *as the top tries to fall over, its pole wobbles in the same sense as its rotation;* right, *as the Moon and Sun try to tip the pole perpendicular to its orbital plane, the pole precesses in the opposite sense.*

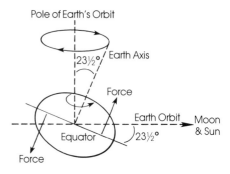

Motion of spin of pole of Encke's comet across the sky. Solid curve represents period of observation. (By Fred L. Whipple and Zdenek Sekanina.)

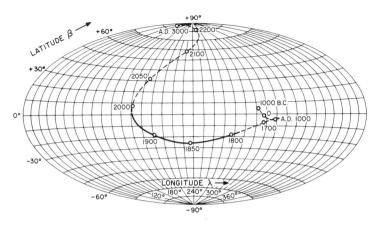

be blown entirely away into space, were raised by the icy vapor. Many of them swirled around to land on the dark southern hemisphere, covering it with a thick layer of meteoric material. Since the 1700s, the south pole has turned around and become well radiated, but not long enough for the insulating layer to be cleared off. As an alternative to our theory, of course, the southern hemisphere may be naturally rocky, because the core of Encke's comet may have been formed that way. The question is a vital one, as we shall see later, because large, old, tame comets may possibly turn into asteroids if their cores are truly rocky. Will we ever know for sure? A space mission to the comet could answer the question.

Sekanina and I predict that, after about 1990, the period of comet Encke will begin to lengthen somewhat during each revolution. But comets are erratic, at best. Perhaps the changing direction of the pole will clean off the rocky debris on one region, or cover up other regions. Nevertheless, the spinning dirty snowball gives a very satisfactory answer to the long-standing question about comet Encke's strange motion.

Chapter 17   **Comet Landscapes**

Because no one has ever seen or obtained a picture of a comet's nucleus, this chapter is liberally sprinkled with imaginative interpretations. Some comets, however, do give us a few clues to the structure of their surfaces. Visual observers have seen and sketched fine, detailed forms around the inner comas near the nuclei of bright and nearby comets. Halley's comet is an excellent example. The narrow jets, fans, and streamers are usually washed out in photographs, although Steven M. Larson at the Lunar and Planetary Laboratory in Tucson is now able to delineate such fine structures from photographs of Halley's comet taken in 1910. He uses subtle techniques of imaging with television-type equipment, called *microprocessing*. He can subtract stray background light, for example, or subtract one image from another to show changes and details. Interpretation, however, remains difficult.

With the 15-inch twin telescopes at Harvard and Pulkovo, George Bond and F. A. T. Winnecke made beautiful drawings of comet Swift-Tuttle 1862 III, the parent of the Perseid meteor stream. Sekanina's analysis of these drawings led him to some surprising conclusions. The jets

Jets from the nucleus of comet Swift-Tuttle of 1862. (Drawn by A. Winnecke at Pulkovo.)

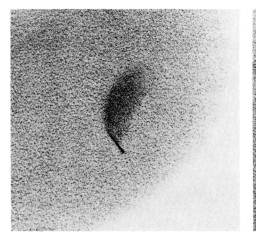
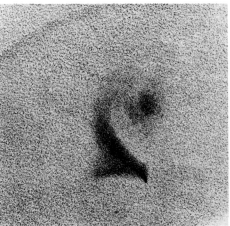

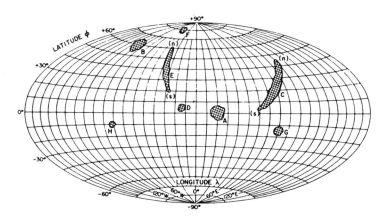

and fans turned out to be dust ejected from several small areas on the surface. He found that the nucleus was spinning rather slowly, in a cycle of sixty-six hours. Each "day" when these small areas came into the afternoon sunlight, they spouted very fine dust in narrow streams. Some of them kept on spouting every afternoon for more than a month, as the nucleus sedately turned around.

Thus Sekanina showed that the surface of a comet is anything but featureless. Its composition probably changes drastically over short distances—hundreds of yards. My speculations took off from here. I reasoned that streams of dust carried by the sublimating ices would expand into wide fans unless the dust spots are surrounded by large areas of vigorous gaseous activity. Widespread outgoing gas is required to constrain the expansion of the dusty jets and keep them narrow. In other words, the dust sources need not be particularly active; they need simply be very dusty, compared with their surroundings, and probably *less* active.

If this is true, the landscape on a comet should show these dusty regions raised above the general terrain, like geological dikes, or perhaps the mesas of the American Southwest. Because a comet has practically no gravity, dust sculptures could attain great heights before being crushed by their weight or by falling over. They might be taller than our highest skyscrapers, perched on odd-shaped, dirty snowballs not very many times greater in size. Bizarre formations of "mushroom" tops, tall slender

A comet landscape as envisioned by astron-omer-artist William Hartmann.

slabs, overhanging ledges, or precipitous gorges possibly may be typical.

Even though this "fantasy land" picture of comets is highly conjectural, a well-observed cometary phenome-non does lend support to this view. Almost all comets are prone to quite unpredictable irregular bursts in bright-ness. At any time, a comet—wild or tame—may brighten by a factor of two to five, or even ten times, usually fading back to "normal" within a few days. The collapse or over-turn of one or more of the hypothetical surface structures in precarious balance provides a reasonable explanation. Sublimation of ices near the base, or some rotational instability might cause the collapse or overturn, which would break up the fragile cometary structure and expose fresh ices, appropriate material for a burst in brightness.

The variations in the brightness of tame comets com-pared with those of wild ones provide another clue to the surface structure and the aging of comets. On the aver-age, with notable exceptions, wild comets brighten and fade away symmetrically about perihelion. Tame comets tend to brighten rapidly just before perihelion and reach their brightest somewhat later; finally, they fade away more slowly than they brighten. This behavior may be partly due to the "lag of the seasons," but there may be other causes: As comets fade on their course away from the Sun, more and more of the larger grains and clods fall back on the surface because the sublimating gas pressure becomes progressively weaker. A thin insulating layer forms on the surface. This layer retards the activity when the comet comes back toward the Sun and must be blown off if the comet is to reach its maximum activity and brightness. Hence, the lag in maximum brightness until after perihelion and the slower fading that follows. As comets grow older, they probably accumulate more and more of this insulative debris and fade away. Whether some of the largest comets become completely clogged with meteoric solids and finally look like asteroids, is, as we have noted, a difficult question to answer.

It is fascinating to speculate about the shapes, landscapes, of split comets. Consider a sungrazer, torn apart by solar tides within the Sun's corona. Are thick layers torn away? Does the nucleus come apart like a broken marble? Does the nucleus have great pinnacles or knobs that the solar tides rip off? We have no answers for these questions about sungrazers, but Sekanina's research on comets that split without apparent cause tells us a great deal about their internal structure and suggests something about the shapes of the pieces.

Most of the split comets can break apart almost anywhere in their orbits, not near the planetary plane where asteroids might bash them, or even very near perihelion. As we have seen, comet West in 1976 was the best performer in many decades. Shortly after a rather close perihelion passage of some 30 million kilometers from the Sun, its nucleus broke into four bright pieces. One secondary piece was observed for about six months. All of the pieces showed the same spectral characteristics. In his notable study of comet West and twenty-one other split comets, Sekanina made three important discoveries.

First, he found that the small pieces deviate more from pure gravitational motion than does the main body of the comet. This deviation suggests that the jet action toward the Sun reduces the Sun's effective attraction on the small pieces more than on the main body, causing them to move in more eccentric orbits. Second, the pieces start to separate at almost zero speed, usually less than a meter per second. Third, the relation between the size (or lifetime) of the pieces and their jet forces away from the Sun is the same for all the split pieces of all the split comets; the smaller the piece, the greater the jet force.

Sekanina thus found that the small broken pieces of comets are just like the large parts they broke away from, that no violent explosion ripped them apart, and that all split comets must be rather similar in their general composition. We must therefore look for the internal disrupting process—a task that will require considerable future study and ingenuity.

Split comets surprise us in another way. For brief periods after they split, the small pieces are brighter than the pieces that last longer and, therefore, must be larger. I think the answer is simple. Material from the inside of a comet, because it is suddenly exposed to sunlight, is free from a mantle of old grains and clods and is partly pulverized. It therefore sublimates faster than the old "weathered" surface. Because any solid body, when broken in two, exposes precisely the same area of fresh material on each piece, the activity of the ices on a comet so broken will be greatest for whichever piece turns its broken side to the Sun. As the pieces spin around, either one can be the brightest. Soon, however, the smaller piece loses a large fraction of its mass because it is smaller and then fades away. Sekanina suspects that the smaller pieces may be layers of the major nucleus, separated like the peel of an orange.

One question about split comets remains puzzling: Why do they split at all? A piece of statistical evidence suggests the disruptive force is internal. Of wild comets in long-period orbits, about 1 in 25 splits during each revolution. Of tamed comets in short-period orbits, only 1 in 170 splits per revolution. Hence, wild new comets are nearly ten times more vulnerable to splitting than old, staid comets. The process may involve exotic ices, which are probably much more prevalent in wild comets. On the other hand, cometary landscapes, or the shapes of comet nuclei, may be even more bizarre than we have dared to imagine. Perhaps comet nuclei are so oddly shaped that very large pieces are unstable. In "falling over," they may really break off and be pushed away by the jet action of the ices exposed near the base.

The suggestion that comets split by spinning too fast has no support from the few cases in which spin periods have been measured. Even so, an extremely long, spindle-shaped snowball, rotating much slower than the critical 3.3-hour breakup period for an ice sphere, might spin off a protuberance by centrifugal force.

Some split comets may really be double or multiple snowballs, rotating around each other in gravitational orbits. The Sun's gravity field could separate them when they come in to small perihelion distances, or the differences between their jet forces might separate them. In this scenario of comet splitting, the pieces would separate slowly without being crushed or having the buried ices exposed. Hence, no bursts in brightness or ejections of large amounts of dust and gas should accompany this type of comet splitting, nor should multiple components appear. A few comets, such as comet West of 1976, show a huge amount of activity as they split. The double-comet explanation can certainly not apply to all (if any) known cases.

In recent years, I have become fascinated with the question of whether any double comets actually exist. Are some of them born as twins, gravitationally coupled to each other in tiny orbits? Astronomers have compared the orbits of comets in the search for groups that suggest a common ancestor. The groups might have been siblings from large comets that split, such as the members of the Kreutz sungrazing family. After a careful study of the statistics of these orbit groups, I concluded—as the Czechoslovakian comet expert, Ľubor Kresák (1982), did later—that the groups came about by chance. There are no more such orbital groups than might be expected from a random sample of orbits. But Kresák and I differed with regard to comet pairs, that is, two different comets moving in similar orbits. He concluded that there were no more pairs than chance would predict, whereas I thought there were a few too many.

Two interesting pairs of comets have added some fuel to the discussion since we published our conclusions on comet groups and pairs. On February 5, 1982, Marc Hartley discovered two comets on a single photographic plate made with the Schmidt telescope at Siding Spring, Australia. The comets were moving in almost identical orbits, nearly in the plane of the Earth's orbit with a period of 5.2 years, but separated by about a week in time. Sekanina concluded, from the orbits and his previous studies of split comets, that a parent comet had split in late 1976 and he predicted that the brighter comet would fade more rapidly than the fainter of the pair because it was moving with greater nongravitational motion. He was correct. By May 1, 1982, the originally brighter component had almost disappeared, having become very much fainter than its companion.

About two weeks after Hartley's discoveries, Syuichi Nakano of Sumoto, Japan, noted that the orbits were similar to the orbit of P/du Toit 2, discovered by D. du Toit at Bloemfontein, South Africa, in 1945 and having a period of 5.28 years, but not recovered since. Marsden calculated that P/du Toit 2 had passed Jupiter at a distance of 0.34 AU in 1963. We are left with two possible conclusions about the comet pair, du Toit-Hartley: The parent was split because of tidal disruption by Jupiter in 1963, or it split in mysterious fashion in about 1976. Tidal splitting at 0.34 AU from Jupiter is highly unlikely because the miss distance of 30 million miles is really very high. On the other hand, a double comet with two nuclei in orbit about each other, might easily have been separated by even this small tidal effect. The question remains: Was comet du Toit-Hartley originally a double comet?

Kresák and three of his colleagues have discovered another intriguing case of a long-lived pair of comets. In 1929, Grigorij Neujmin of Simeis, Ukrainia, discovered a comet with a 10.9-year orbit moving very close to the Earth's orbital plane. It was faint and recovered only in 1951 and 1972. In 1954, George Van Biesbroeck, the indefatigable comet observer at Yerkes Observatory in Wis-

George Van Biesbroeck at the lens end of the 40-inch Great Refractor, circa 1956. (Courtesy Yerkes Observatory, University of Chicago.)

consin, discovered a comet of medium brightness in a vaguely similar orbit with an orbital period of 12.4 years; it was recovered in 1966 and 1978. Tracking back these comet orbits, Kresák and his colleagues calculated that both encountered Jupiter on March 10, 1845, at a distance of only 0.04 AU and found that the preencounters were identical within the accuracy of their calculations (less that 0.2 percent in orbit, shape, and size; less that 0.6 percent in angles). Were these two rather healthy comets, which have survived for 140 years after their Jupiter adventure, originally a single or a double comet? If single, it may have split, or if double, it may have been separated by Jupiter's tidal effect in 1845. Either alternative is possible.

Even more interesting is the strange, perhaps unique, activity of a comet discovered by the British astronomer, E. Holmes, on November 6, 1892. It was a member of Jupiter's family with an orbital period of 6.9 years and perihelion distance of 2.1 AU. The great comet observer, Barnard, was then at Lick Observatory. He had already

Comet P/Holmes near the Great Andromeda nebula as photographed by Edward E. Barnard on November 9.2, 1982, shortly after discovery. (Courtesy Lick Observatory.)

discovered fifteen comets and observed many more. Thus his description of P/Holmes on November 9 is highly significant: "Its appearance was absolutely different from any comet I have ever seen—a perfectly circular and clean cut disk of dense light, almost planetary in outline with a faint, hazy nucleus . . . (brightness = Andromeda Nebula)." By the next night, it brightened perceptibly and he saw an outer faint diffused envelope some 80,000 kilometers in diameter.

The comet must have brightened about a hundred times within a very few days before discovery. It was ideally placed for observation in the northern sky, not far from the frequently observed Andromeda nebula, and should have been discovered earlier unless it had been much fainter. What distinguished P/Holmes besides its unique appearance was its rare variation in brightness. It faded very little for nearly a month, its coma growing larger all the time. Then it plummeted in brightness by perhaps 200 times. By January 15, 1893, it looked like a faint globular cluster. On January 16, observers in Europe were astonished to find that the comet had almost regained its original naked-eye brilliance. It then faded quickly and was last seen in 1893 during April.

When P/Holmes was recovered in 1899 and 1906 it was

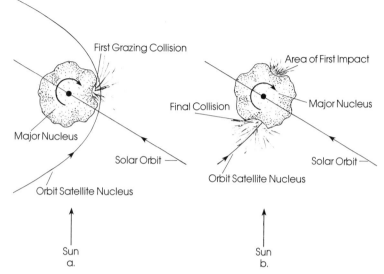

Scenario of encounter between satellite and nucleus of Holmes comet: a, grazing collision; b, final impact, seventy-three days later.

First Grazing Collision

Area of First Impact

Final Collision

Major Nucleus

Major Nucleus

Solar Orbit

Solar Orbit

Orbit Satellite Nucleus

Orbit Satellite Nucleus

Sun
a.

Sun
b.

inactive, intrinsically nearly 10,000 times fainter than in its moment of glory in 1892. No one detected it again until 1964, but it was photographed in 1972 and 1979, having lost another factor of ten or so in brightness.

In 1981, after Thomas C. Van Flandern had revived the question of possible comet twins, I looked into the theory of such orbital pairs and collected the many observations of P/Holmes in 1892–93. I found, theoretically, that if the nongravitational jet action is greater for one than for the other, their mutual orbit can twist around so that it becomes more and more elongated. The pair can finally collide.

In the case of P/Holmes, the first collision could have been a grazing encounter as the two nuclei spiraled together. On the next encounter, seventy-three days later, the collision could have been more central and final. From measures of expanding shells near the nucleus, I found that P/Holmes was rotating with a period of 16.3 hours, unchanged from the middle of November 1892 until well past its second outburst in 1893. After the first outburst, only one area on the nucleus was active. A second one, however, appeared after the second outburst. The phases of the active areas fitted the theory of a collision scenario if the unobservable geometry of the orbit was correct. In this scenario, the grazing collision set off a very active region on the major nucleus, and the final impact created a second active area while reactivating the first one.

Hence, P/Holmes may well represent the collisional demise of a small satellite comet in orbit about a larger nucleus. But we may never know for certain. Only one other comet has acted like P/Holmes. This is the faint comet P/Tuttle-Giacobini-Kresák, a member of the Jupiter family with a period of 5.6 years. In 1973, it exhibited two gross outbursts of some 4,000 times in brightness, separated by about forty days. Since then, it has been very faint and inactive. It, too, may have been a double comet in which the larger of the twins cannibalized its sibling.

Whatever the true cause of comet splitting, and whether or not some comets are double, evidence abounds that our dirty snowballs may have shapes other than round spheres. Their surfaces are clearly not uniform in composition, even over small distances. Some areas seem to be very active and some very dusty but still active; still others are probably covered with rocky meteoric material. As the ices sublimate from more active icy regions, they may leave behind large grotesque formations. The Space Age gives us hope that someday we may see televised pictures of comet nuclei and thus be saved from further speculation, at least about comet landscapes. This hope is bolstered by the flood of new knowledge about comets that the Space Age has already given us.

Chapter 18   Comets in the Space Age

Until the Space Age, astronomers had truly seen the universe "through a glass darkly." The human eye detects light over only one octave out of hundreds in the electromagnetic spectrum. The atmosphere frugally allows heat or infrared radiation to pass through only a few "cracks." After World War II, radio extended our range of detection to much longer wavelengths, greatly expanding our knowledge of planets, stars, galaxies, and the Universe, but not of comets. As discussed earlier, the atmosphere is an opaque curtain to ultraviolet light, precisely the region in which the atoms and molecules in comets radiate their most instructive spectral "fingerprints." The first comet sensed in the far-ultraviolet from space—a bright comet discovered by the Japanese triumvirate Akihiko Tago, Yasuo Sato, and Kozo Kosaka in 1969—displayed a collosal cloud of hydrogen around it. Other bright comets, observed from space in the far-ultraviolet, showed similar clouds. For the first time, we could measure *accurately* the huge amount of gas lost by great comets, thanks to both the rockets and instruments of the Space Age.

The detection of the oxygen atom, radiating also in the far-ultraviolet, told us more about the composition of the gas. Although we already had rough estimates of the rates at which comets lose OH molecules—achieved by difficult measurements in the near-ultraviolet where the Earth's atmosphere grudgingly transmits only a small fraction of the light—the answers were uncertain. From space, all the components of the water in comets—hydrogen, oxygen, and OH—could be measured accurately. The observed numbers added up very nearly to two of hydrogen for one of oxygen—that is, water. These measures have held up for recent comets observed in space, and thus indicate that other possible molecules, such as methane (CH_4) or ammonia (NH_3), can't contribute very much hydrogen.

The much-maligned comet Kohoutek of 1973–74, observed so thoroughly from the ground and from space,

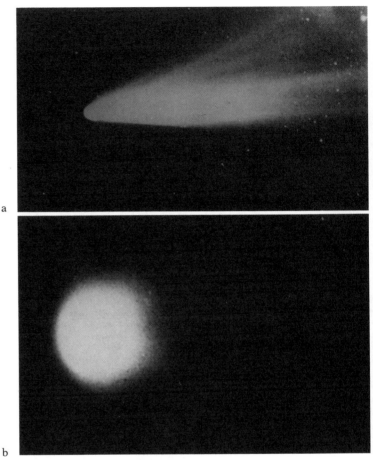

a

b

Photographs of Comet West 1976 VI, obtained from a rocket on March 5, 1976, and printed to the same scale: a, *visual light photograph (courtesy Paul D. Feldman);* b, *Lyman alpha photograph showing the hydrogen cloud in ultraviolet light (courtesy C. B. Opal and G. R. Carruthers, U.S. Naval Research Laboratory).*

provided a gold mine of new facts about the nature of comets. Comet Kohoutek is the wild comet we discussed that was spotted as it became active far from the Sun. Although predictions that it would brighten by 25 million times at perihelion and become spectacular were over-optimistic, it did brighten by a million times and become visible to the naked eye. Even though the comet was a disappointment to the public, it was a great success of modern comet history for scientists. Not only did it show a great hydrogen cloud, as it lost a million tons of water vapor per day, but its ultraviolet spectrum brought out carbon lines as well as the oxygen lines never before observed. When the Earth crossed its orbital plane, the comet showed a strong forward, or sunward, tail of grains and clods. E. G. Gibson, the astronaut inside Skylab, drew sketches of the comet's head and tails in December 1973 and January 1974. Large particles making up the sunward tail had been blown out weeks before and were moving away from the comet, close to the orbital plane.

Comet Kohoutek exhibited a dust tail that was a fine sight both from the ground and from space. Edward Ney, with his 30-inch telescope at the University of Minnesota, detected infrared absorption from silicates in these tiny tail particles. This was the first direct proof that silicon exists in comet grains, confirming the indirect evidence in the spectra of meteors.

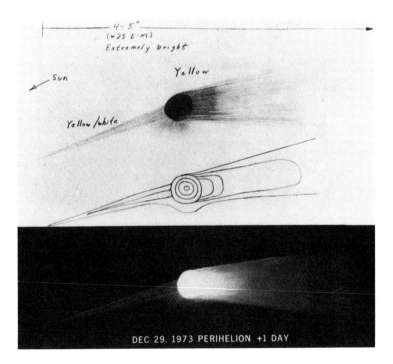

Drawings of comet Kohoutek of 1973 from the Skylab satellite, showing the sunward tail when the Earth was near the plane of the comet's orbit. (Courtesy E. G. Gibson, National Aeronautics and Space Administration.)

With comet Kohoutek, radio made its debut as a powerful tool for observing comets. B. L. Ulrich and E. K. Conklin used the 11-meter microwave disk at Kitt Peak, Arizona, to detect the most complex molecule yet found in comets, methyl cyanide (CH_3CN). They were quickly followed by David Buhl, W. F. Huebner, and L. E. Snyder, who used the same radio telescope to discover lines from hydrogen cyanide (HCN). These molecules are well-known in the great interstellar clouds. Their presence in comets is therefore very suggestive with regard to cometary origins. At much longer wavelengths, CH and OH were found for the first time by radio. A few months later, W. M. Jackson, T. Clark, and Bertram Donn, using the Haystack antenna of the Massachusetts Institute of Technology, found a centimeter-wavelength line of water in the bright comet Bradfield (1974 III).

Comet Kohoutek stimulated Susan Wyckoff and Peter Wehinger in Israel to track down the source of a number of hitherto unidentified red spectral lines, which they had observed in the comet. With the help of the spectroscopist Gerhard Herzberg (a Nobel Prize winner) in Canada and the Lick Observatory astronomer George Herbig, they identified the presence of water by means of these red lines coming from the positively charged ion. At last we had proof, nearly 200 years after Laplace's brilliant suggestion, that indeed water vapor is present in comets.

Ernest Hildner and his colleagues of the High Altitude Observatory in Boulder made a motion picture of comet Kohoutek using 1,600 still photographs taken from Skylab—the first comet movie, not just a few photographs combined to show motion.

The power of modern radar finally expressed itself by giving direct proof of a sizable, discrete nucleus in a comet—none other than our favorite source of cometary knowledge, P/Encke. This happened in 1980, when P. G. Kamoun and his colleagues detected radar echoes from the comet with the great 1,000-foot-diameter dish at Arecibo, Puerto Rico. Encke's comet was 0.33 AU away from the Earth at the time, or 50 million kilometers, so that radar detection was a truly remarkable feat. The scientists lacked knowledge of the reflecting power of the unknown surface at the radio wavelength of 13 centi-

Aerial veiw of the 1,000-foot radio dish of the Arecibo Observatory in Puerto Rico. (Courtesy Arecibo Observatory, and Ionosphere Center, Cornell University.)

Comet IRAS-Araki-Alcock 1983d as photographed by Antonín Mrkos in Czechoslovakia on May 10.0, near its closest approach.

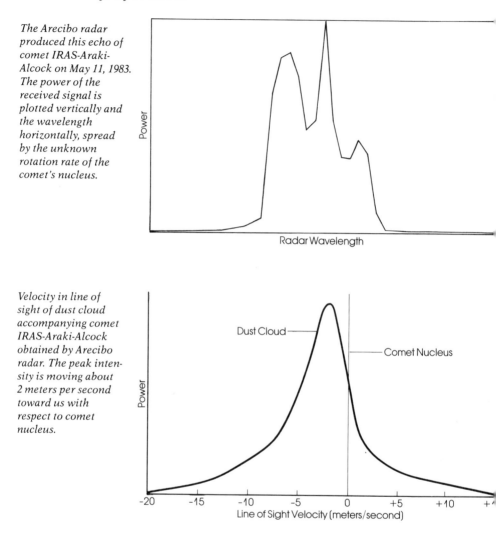

The Arecibo radar produced this echo of comet IRAS-Araki-Alcock on May 11, 1983. The power of the received signal is plotted vertically and the wavelength horizontally, spread by the unknown rotation rate of the comet's nucleus.

Power

Radar Wavelength

Velocity in line of sight of dust cloud accompanying comet IRAS-Araki-Alcock obtained by Arecibo radar. The peak intensity is moving about 2 meters per second toward us with respect to comet nucleus.

Dust Cloud

Comet Nucleus

Power

-20 -15 -10 -5 0 +5 +10 +

Line of Sight Velocity (meters/second)

meters and therefore could not make a reliable calculation of P/Encke's diameter, which may be 2 kilometers or so. But a nucleus was definitely present! The dirty snowball was no longer just a theory; it had become an observed entity.

The unexpected comet IRAS-Araki-Alcock (IAA) came so close to the Earth in May 1983—it was only 4.7 million kilometers away—that it was an easy radar target. Both the great Arecibo dish and the Goldstone 210-foot dish, operated by the California Institute of Technology, detected it easily, but they had no time to plan thorough programs of observation. Even though the huge radar echoes obtained by both dishes did not give us an image of a comet landscape, they did, indeed, show that the nucleus is irregular in shape. No precise measurements of the diameter of comet IAA have yet been published, but it seems to be in the same size range as P/Encke's, but may be somewhat larger, perhaps a few kilometers across.

Comet IAA created a radar surprise! A strong echo appeared that was not moving at the same speed as the nucleus. Apparently the radar picked up echoes from clods, centimeters to meters in diameter, that the nucleus had blown off gently in recent weeks before the observations. Undoubtedly this lost mass forms the sunward tail seen optically and accounts for some of the heat radiation measured by the infrared-sensing satellite, IRAS. It is ironic to find that comets support actual flying sandbanks and gravelbanks, but not as a part of their nuclei. They throw off this flying rubble, which is then lost to the comet forever. The radar demonstrates just how meteor

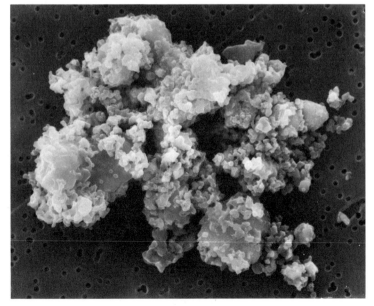

Electron microscope picture of stratospheric particle thought to be a piece of a comet. The total width of the picture represents less than 1/1,000 inch. (Courtesy Donald E. Brownlee.)

streams form, streams that become meteor showers if the Earth passes through them.

With regard to the meteors, the modern high-flying aircraft and stratosphere balloons have made possible a new method of detecting some of these tiny particles. For the past decade, Donald E. Brownlee, of the University of Washington at Seattle, has studied the fine dust collected by these upper-atmospheric vehicles. Some of the particles are unique, unlike any man-made or natural particles. Long ago, Öpik and I showed that, under special circumstances, very small meteor particles can be stopped by the atmosphere without being boiled away. They must come in at low speeds and travel almost tangentially to the atmosphere in order to slow down before they reach the denser atmosphere below. Most scientists now believe that the "Brownlee" particles are samples of comets—the only samples that have not been thoroughly melted. Some deep-sea spherules are probably melted grains from comets. Many of the Brownlee particles, when imaged with the electron microscope, look like clusters of fish roe or grapes, with individual eggs or grapes some 0.00025 millimeters (1/100,000 inches) in diameter. They are another vital clue to the nature and origin of comets.

Thus we see that the Space Age has lifted the astronomer and his telescope to a new vista and provided him with subtle sensing instruments, some not even dreamed of a century ago. Radar has proved that the heart of a comet is a solid snowball, and radio has unveiled secret new molecules, particularly methyl cyanide and hydrogen cyanide, previously detected only in great interstellar clouds. Radar and infrared instruments have measured the cometary rubble that makes up meteor streams. Ultraviolet light detectors outside the atmosphere have given the first quantitative measurements of the broken components of water, showing that, indeed, water is the principal ice in our dirty snowballs. Other measures of radiation in the radio, infrared, and ultraviolet regions of the spectrum have added to our knowledge of comet constitution and structure. Finally, sensors of the solar wind have clarified many of the strange activities in the great tails of comets.

Chapter 19 The Elemental Composition of Comets

The next step in understanding comets and their origin is to consolidate the clues about their composition that Space Age science and technology have given us. These clues tell us more about the distribution of the atomic elements in comets than about the chemical compounds, even for the ices. We need a standard for comparison, and the obvious standard is the Sun. All our evidence suggests that the Earth, meteorites, the Moon, and the Sun evolved from the same materials, and that the various separation and collection processes controlled the end products. If the composition of comets turns out to be consistent with their having been frozen out of a solar mix of elements, then comets *may* have originated at the same time as the Solar System. What composition should we then expect for comets?

Both the dirty snowball model and comet activity show clearly that comets must have been frozen out at very low temperatures, well below that required to freeze water-ice. Otherwise, comets could not be so active at solar distances beyond about 2 AU. More volatile ices than water-ice must be present in comets. Even at extremely low temperatures near absolute zero, hydrogen cannot freeze in space, nor can the noble gases (which form no compounds), helium, neon, argon, krypton, and xenon. If comets were made from a solar mix of elements, we could expect to find compounds that will freeze at perhaps 50° to 100° or less above absolute zero, plus all the heavy elements like silicon and iron that freeze at very much higher temperatures. The noble gases should be absent because they freeze at such low temperatures and do not form compounds. With these constraints in mind, we can determine whether comets might be composed of a solar mix of materials.

According to A. G. W. Cameron of the Harvard-Smithsonian Center for Astrophysics, 77.5 percent of the Sun is hydrogen and 20.8 percent helium—these two principal components add up to 98.3 percent of the Sun's total mass. Of the remainder, from which we might make comets, 1.3 percent consists of carbon, nitrogen, and oxygen and about 0.4 percent of the heavier elements to make the minerals in the grains. A tiny fraction of the hydrogen will combine with the carbon, nitrogen, and oxygen to make ices, and about one-third of the oxygen will be locked up in the minerals of the grains. These assumptions lead to the conclusion that the ices of comets should contain about 37 percent carbon, 9 percent nitrogen, and 54 percent oxygen by weight, supposing they were frozen from the solar mix of elements. The minerals in our dirty snowball should add up to about 37 percent of the total mass, or about 60 percent of the mass of the ices. Now let us see how well the observations confirm our expectations.

This question is not as easy to answer as it might seem, even with the vital knowledge we have gained from the spectra and the magic tools of space. The problem arises in the comet itself, because of events that occur in the first minutes after the icy vapor lifts off from the nucleus. During this short time, the atoms and molecules, or shreds of molecules, bounce around against each other, as they do in ordinary gas. Furthermore, they can collide with the grains, some of which may be adding ice as they sublimate. All this time, they are affected by sunlight. As a result, the inner coma of a comet is a chemical factory! This leaves us confused as to whether the materials we detect come unchanged directly from the nucleus or were manufactured near the surface. Fortunately, the tools for analyzing this horrendously complicated problem have become available in recent decades.

Walter F. Huebner, at the Los Alamos National Laboratory, and other chemists attack the problem with huge computers. These powerful number crunchers can keep track of more than a hundred species of atoms and molecules and a thousand and more possible reactions among them. Included are ordinary gas-phase chemical reactions and also reactions induced by the solar photons. Huebner lists fifteen different *types* of such reactions, which separate or combine these hundred or so species. For example, an ultraviolet quantum may strike any neutral atom or molecule and transform it into an ion, the electron being knocked out. The electron may strike another molecule such as nitrogen, N_2, breaking it up into two atoms, N and N, and so on. Thus, neutral molecules may be broken up or ionized and new ones formed. From our vantage point on Earth, which is so distant from comets, we can observe only the end products of the chemical factory after they have escaped hundreds or thousands of kilometers into space, where the gas is so rare that collisions no longer count. Thus the complicated gas-phase chemistry disguises the composition of the original ices in a comet. Suppose that some possible basic molecule, such as methane (CH_4), lifts off from the surface. By the time it goes through the maelstrom of the gas-phase chemistry, it probably has been broken up, and we would observe CH, C, H_2, or the like. To make matters even worse, suppose that we started with several basic elements such as hydrogen, oxygen, nitrogen, and carbon. They may combine into CO, NH_3, OH, and so on, or even into water, cyanogen (CN), or methane (CH_4).

The problem of ascertaining the original icy composition of the comet is thus truly horrendous. Giant computers, luckily for us, can remember all these processes and calculate the probability of each reaction, the basis being laboratory measurements made by many chemists over decades. Putting all these calculations together with the latest observations of comets, chemists can make increasingly better judgments concerning the actual ices in comets. Let us now look at the data from current observations.

Water appears to be very abundant. Our basic molecules of ammonia and probably methane have now been detected, but not in large quantities. Carbon monoxide (CO) and dioxide (as CO^+), as well as C_2 and C_3 are all seen. Sulphur was expected, and finally found, both as an atom, combined with carbon as CS, and in the unlikely laboratory form of the double molecule, S_2. Interestingly, hydrogen sulphide (the molecule that smells like rotten egg) shows up as an ion, H_2S^+, in the tails of recent bright comets.

We have noted the radio contribution of the deadly gases hydrogen cyanide (HCN), and methyl cyanide, (CH_3CN). The Earth-grazing comet IRAS-Araki-Alcock in

1983 not only gave us the sulphur molecule S_2, but probably also the long-awaited formaldehyde (H_2CO) molecule and certainly its broken daughter, HCO. Thus CO, broken from HCO by sunlight, may be a source of the carbon monoxide seen in comets, because carbon monoxide is very difficult to "freeze out" in space, even at temperatures near absolute zero.

Among the heavier atoms, potassium, calcium, and several metals (iron, cobalt, nickel, and copper) show up in sungrazers. Sodium is so easy to excite that, as we mentioned earlier, it can be seen in comets out to nearly 1 AU from the Sun. As expected, the noble gases such as helium, neon, argon, or krypton have not been detected. Not only are they practically impossible to "freeze out," but their spectra are not easy to observe. Even if some of their atoms are trapped in the ices, we have little chance to find them.

What do all these clues tell us about the ices? The world expert on this subject is the astronomer Armand H.

Armand Delsemme.

Gassy comet Tago-Sato-Kosaka 1969 IX. (Courtesy H. J. Potter and A. N. Sokalov, Soviet Expedition to Chile.)

Delsemme, who began his comet investigations in his native Belgium. After exciting adventures in the Belgian underground during World War II, he finally settled down at the University of Toledo, Ohio, to continue his study of comets. He concludes that water-ice is by far the most abundant ice in comet nuclei. In his synthesis, Delsemme needs all of the hydrogen observed to account for the water, combining it with the observed O and OH in bright comets. Only the errors in the measurements and in his calculations leave any hydrogen to combine with carbon and nitrogen. This apparent shortage of hydrogen is not worrisome, in view of the uncertainties in the measurements and theory, and the small amount of carbon that may have been combined with hydrogen. All in all, cyanides of hydrogen and methyl, the formaldehyde, and the cyanogen constitute only about 2 percent of all the ices.

Delsemme finds that the amount of nitrogen, compared with oxygen, is close to what was expected, within an uncertainty factor of less than two. Carbon, however, is lacking by a factor of three to four times, a definitely unexpected result. Possibly the carbon is caught in the grains, which are rather black, but we cannot tell yet how much carbon the grains actually contain. Alternative scenarios will be discussed later when we look more deeply into methods of making comets. Sulphur is almost exactly as abundant as expected, about 1/40th the mass of oxygen.

The dust-to-ice ratio in comets is difficult to estimate. Comets are so diverse! Some, as we have seen, appear to be extremely dusty. Examples are comet Arend-Roland of 1957, the great comet 1910 I that upstaged Halley's comet, and the split comet West of 1976. On the other hand, our old friend P/Encke, comet Tago-Sato-Kosaka, comet Whipple-Fedtke-Tevzadze (1943 I), and many others show almost a purely gaseous spectrum. Halley's comet falls between these extremes. We probably miss seeing most of the sizable grains and clods from comets because the tiny grains are very effective in scattering sunlight, in comparison with the larger ones. Thus comets that appear to be purely gaseous may actually be losing many sizable pieces that do not scatter light well. More infrared observations may answer the question by measuring the heat from the totality of grains. For the moment, we can only conclude that the observed dust/gas ratio rarely exceeds 1.0. The average expected value was 0.6; if a considerable fraction of the carbon lies in the grains, however, a higher value would be predicted.

Because a few comets show very dusty regions on their surfaces, it seems reasonable to suspect that some comets really are dustier than others. This may be an important clue to comet formation, evidence that comets formed in quite different celestial environments—some with an

A very dusty comet, Mrkos 1957 V, as photographed by Alan McClure.

abundance of very fine dust grains and others with almost pure gas or ice.

The observed composition of cometary grains and clods seems to hold no surprises. Peter M. Millman of Canada, a meteor expert for half a century, finds from meteor spectra that this cometary rubble shows the expected abundance ratios of sodium, magnesium, calcium, and iron. Gale A. Harvey at NASA's Langley Research Center adds manganese, nickel, chromium, silicon, aluminum, and oxygen to Millman's list of elements observed in meteors with the expected abundance ratios. Brownlee's upper-atmospheric particles, which probably come from comets, confirm these conclusions.

Spectrum of a Perseid meteor from ultraviolet (0.4 microns) to near-infrared (0.9 microns). (Courtesy Dominion Observatory, Ottawa, Canada.)

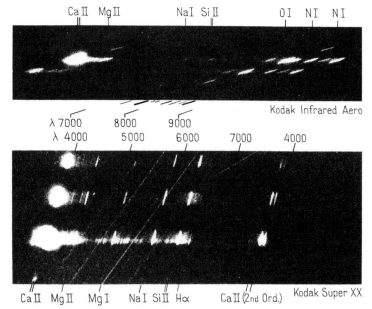

Spectrum of a Perseid meteor. (Courtesy Harvard College Observatory.)

Only a rendezvous mission to a comet can answer satisfactorily the question of its composition. A sample-return mission, analysis at the spacecraft, or analysis with a penetrating probe will be needed. In the meantime, I am happy that we have learned as much as we have about comets from observations at great distances. Their composition is about what we should expect if they had been frozen at very low temperatures from a mix of elements like the Sun's. Carbon is somewhat deficient in the ices, but it may be hidden in the dust. We know that water-ice is very abundant, but are still ignorant of the other chemical compounds that make up the ices of comets. To continue this story, we must go to the laboratory to find out what kinds of strange ices our chemists can make—some that may have been frozen in comets.

Chapter 20

Exotic Ices in Comets

If water-ice were the only active ingredient in comets, we should expect them to begin to brighten as they come within about 2 to 3 AU from the Sun. On receding, they should fade away a bit farther out. The constituents could include some other compounds of carbon, nitrogen, and hydrogen, mixed with grains of heavier elements. But many comets, as we have seen, cannot be explained so readily. They busily send off gas and dust at solar distances where the temperature is so low that ordinary water-ice is quite dormant, even when it may have trapped volatile ammonia or methane molecules as clathrates.

To help us out, chemists tell us that the ice we see in winter ponds and in our drinks is not the same as the ice they can make by freezing out water vapor at very low temperatures. According to Roman Smoluchowski at the University of Texas, the molecules in specially manufactured ice cannot arrange themselves into a perfect crystalline structure; instead, all confused, they make an amorphous ice without a pretty mathematical arrangement. If frozen at temperatures below 10° Kelvin (−440° F), the amorphous ice is about 17 percent less dense than ordinary ice, and if frozen just below 130° K (−227°F), it has about the same density as ordinary ice. When amorphous ice is warmed up above 153° K (−185° F), the molecules rearrange themselves into the usual crystalline form and, in so doing, give off heat.

Amorphous ice thus appears to be a convenient solution to our problem. Smoluchowski expects the surface of a comet to be warmed to 153° K at perhaps 5 to 7 AU from the Sun if the nucleus is spinning rather slowly—this temperature is warm enough to activate amorphous ice and initiate comet activity. An actual comet, however, could not have been formed from pure water vapor under ideal laboratory conditions. It must have grown from tiny grains or snowflakes, pummeled all the while by many types of atoms and molecules, not to mention occasional

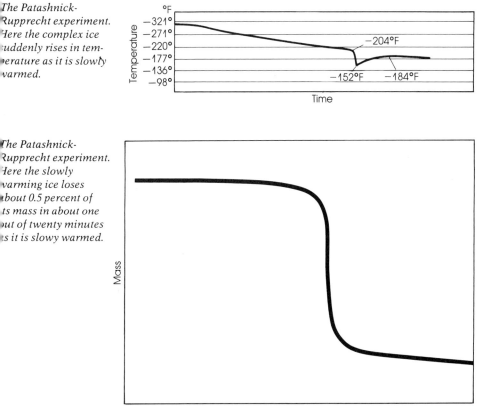

The Patashnick-Rupprecht experiment. Here the complex ice suddenly rises in temperature as it is slowly warmed.

The Patashnick-Rupprecht experiment. Here the slowly warming ice loses about 0.5 percent of its mass in about one out of twenty minutes as it is slowy warmed.

light quanta and cosmic rays, which must also have messed up the chemistry. With this in mind, Harvey Patashnick and Georg Rupprecht at the Dudley Observatory, Albany, New York, set up some sophisticated laboratory experiments. They froze out water vapor mixed with carbon dioxide, methanol (CH_3OH), or ammonia at liquid-nitrogen temperatures (77° K or −321° F). Their report to NASA in 1977 showed that, when they warmed up these mixed ices to about 139° K (−210° F), the ices spontaneously heated by several tens of degrees and blew out about 1 percent of their mass into the vacuum tank in which they were frozen. These amorphous ice mixtures acted much like comets coming in toward the Sun.

Assuming that the ices in comets really are amorphous,

or at least complicated mixtures, Bertram Donn at NASA's Goddard Space Flight Center in Maryland decided to find out by experiment what cosmic rays would do to them. Remember the conjectures that wild comets such as Kohoutek tend to brighten up much farther from the Sun than tame comets, possibly because of an outer layer that was damaged by cosmic rays. Marla H. Moore, at the nearby University of Maryland, extended Donn's early experiments into a larger research project, her doctoral thesis. On a cold surface at the low temperature of only $20°$ K $(-424°$ F$)$, she froze out water vapor mixed with various gases such as ammonia, methane, and nitrogen. Then she radiated the thin ice layers with million-volt protons (hydrogen nuclei) from a Van de Graaff accelerator, subjecting them to conditions similar to the bombardment by soft cosmic rays expected in an interstellar cloud during 4.6 billion years, the age of the Solar System.

Her results were startling. When only slightly warmed by $10°$ or $15°$, the damaged ices began to fluoresce and, in some areas, were close to exploding, shooting out flashes of light and gases in profusion. When warmed by another $100°$ and more, the irradiated ices again fluoresced until the temperature reached about $150°$ K $(-190°$ F$)$; above this temperature the activity stopped. Clearly, the second activity at about $150°$ K was connected with the change from amorphous ice to ordinary ice. At the very low temperatures, the molecules in the ices had been truly damaged, in the sense that high-energy protons had knocked about the electrons and even atoms, displacing them from their frozen positions within the molecular latticework. A small amount of heat provided enough energy to enable many of the molecules to repair themselves and release some trapped energy and light in the process.

Naturally, Moore wanted to find out what strange compounds had been made by the violent pummeling from the high-energy protons. She found, indeed, that although the ordinary ices were quite amorphous, many new molecules had been synthesized, such as C_2H_4, C_2H_6, CO, CO_2, and nitrogen compounds. None of these molecules occurred in the original gases. The cold ices had turned a bit yellowish, and, when finally warmed to room temperature and pressure, left a residue of yellowish-brown

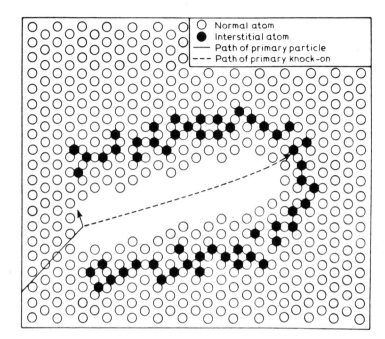

Damage to crystal structure by cosmic-ray particle entering at lower left and knocking on an embedded nucleus, as envisaged by J. A. Brinkman.

"gunk." The gunk, amounting to about 1 percent of the ices, was by and large water soluble, but too complicated for easy chemical analysis. It appeared to be made of quite complex hydrocarbons and nitrogen compounds when nitrogen or ammonia were originally present, one derived compound being ammonium azide, NH_4N_3.

Moore's experiments strongly support the idea that wild comets, after billions of years of exposure to cosmic rays in outer space, develop an outer layer, perhaps a meter thick, of highly damaged ices. On first coming in from the Öpik-Oort cloud, the outer layer becomes active at great solar distances because, for the first time, the comet has been warmed a bit by sunlight. After the outer layer is sublimated away, the underlying layers will be typical of ordinary comets that have traveled near the Sun before.

A few comets, like comet Kohoutek, which brighten at unusually great distances from the Sun, strongly support the cosmic-ray theory as described above, coupled with the assumption that these comets were indeed formed billions of years ago, perhaps with the Solar System.

Another set of clues supports this idea. Brian Marsden noted in 1976 that most of the comets with large perihelia, around 3 to 4 AU from the Sun, are wild comets having nearly parabolic orbits. Many more should therefore be known to have returned one or more times. He suggested that, on subsequent returns, the comet is fainter than on its first approach to the Sun, too faint to be discovered, and thus few comets of this type have been observed.

Edgar Everhart shows that there are many fewer longer-period tame comets among those with smaller perihelion distances than one would expect. This conclusion along with Marsden's can be explained by an alternative theory: At present there is a "bull market" in wild comets. We shall discuss later the important problem of possible causes of a burst of wild comets. Whatever explanation may be valid, the laboratory shows that cosmic rays could damage the outer layer of comets and cause them to brighten abnormally when they first approach the Sun.

But what about the other exceptional activities of comets, in particular, that of our anomalous comet Schwassmann-Wachmann 1 (P/SW1), which bursts majestically from time to time while remaining just outside Jupiter's orbit? Are amorphous ices able to explain this comet's activity, and that of many others, at solar distances where ordinary water-ice is dormant? To pursue this question further, we must consider a more detailed process of comet formation. And for this we must look to interstellar gas and dust clouds as the ultimate source from which comets, as well as the Sun and the entire Solar System, were formed.

The Brownlee particles caught in the Earth's upper atmosphere are almost certainly pieces of comets. The fact that they consist of extremely tiny particles suggests that they may have accumulated from interstellar dust grains. Fortunately, Mayo Greenberg, at the Leiden Observatory in the Netherlands, has spent several years conducting laboratory experiments that show just how interstellar dust grains can grow. He relates his experiments and methods to the many astronomical observations of interstellar dust clouds, to the gas within them, to

COMET SCHWASSMANN-WACHMANN I

1961   Oct. 12   Oct. 18   Nov. 3

Outburst of comet P/Schwassmann-Wachmann 1 in 1961 when comet was well beyond Jupiter's distance from the Sun. Note development of spiral structure caused by slow rotation of nucleus, called the "ring-tailed snorter" by observer, E. Roemer. (Official U.S. Navy photograph.)

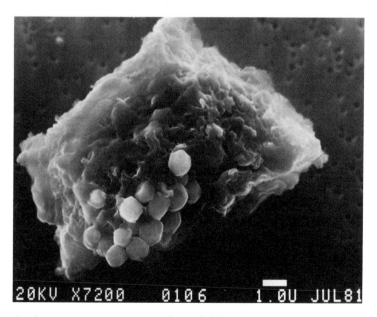

An electron microscope image of a probable cometary particle. The mark (lower right) is 1 micron, or 1/1,000 millimeter, in length, or 1/25,400 inch. (Courtesy D. E. Brownlee.)

Negative photograph of star-forming gas-dust cloud in Orion lighted by red line of hydrogen. (Courtesy Yerkes Observatory, University of Chicago.)

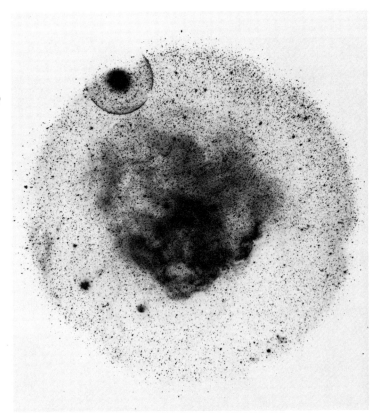

the new stars that blow off as novae or supernovae, and to the radiation that falls on dust grains in space.

Distant stars seen through dust clouds appear redder than nearby stars of the same intrinsic color. The dust grains scatter blue and ultraviolet light more than yellow or red light because the average diameter of the grains is about one-sixth the wavelength of red light. The longer red waves scarcely notice the tiny dust grains, whereas the shorter waves of blue light are more nearly the size of the grains and are more scattered by them. Studies of the reddening by fine particles have provided some information about the conditions under which interstellar grains form. Greenberg simulates these conditions in his laboratory. Temperatures in space are very low, only a few degrees above absolute zero, while faint starlight seeps through the clouds. Although cosmic rays are present, they strike the grains for only millions to hundreds of millions of years, instead of the billions of years that are required to damage the outer layers of comets. Hence, we can disregard them here.

To simulate the making of interstellar ice grains, Greenberg freezes out likely gases at about 20° K (−424° F), while irradiating them with ultraviolet light similar to typical dim starlight. He collects the ices on a "cold finger" inside this apparatus. On warming the cold finger upon which the ices have formed, he observes tiny but violent explosions and light flashes, similar to the ones noted by the experimenters mentioned above. Finally, at room temperature, he is left with a small percentage of residual yellow material composed of complex hydrocarbons and other compounds. Its chemical analysis is still incomplete, but it certainly contains many complex compounds.

The effect of the ultraviolet radiation on the ices is startling. When Greenberg starts with the gases carbon monoxide, ammonia, water, and methane, and does not radiate them, only the original compounds show in the absorption spectrum of the ices. With radiation, he finds others in addition: carbon dioxide (CO_2), formaldehyde (H_2CO), the radicals HCO and CH_3, and more complex hydrocarbons that have not yet been fully identified. His experiment makes ices that sublimate at much lower temperatures than does water-ice and that give out some of the gases actually observed in comets. I can call this unknown material in comets *complex ice*, meaning an amorphous water-ice mixed with a large variety of other compounds and radicals made from carbon, oxygen, nitrogen, and hydrogen, as well as traces of almost all the known elements, except possibly the noble gases. The composition of complex ice, so defined and formed at extremely low temperatures, is consistent—as far as we know—with the composition of comets. The various laboratory experiments indicate that complex ice may be almost explosive when warmed. After it has reached temperatures above about 153° K (−185° F), complex ice may revert to ordinary water-ice, which has become contaminated with many other compounds and elements.

If we now admit complex ice as a principal component

of comets, we have a basis for understanding the strange behavior of such comets as P/SW1, bursting out at 6 or 7 AU from the Sun. Long ago, Charles A. Whitney at the Harvard Observatory suggested that the activity of P/SW1 demands a surface energy storage mechanism to collect solar heat for perhaps many months. Suppose, following a suggestion by Smoluchowski and others, that complex ice is slightly warmed under a thin, partly insulating layer of water-ice and stony grains. The complex ice over a large area, perhaps a square kilometer or so, has become almost warm enough to change state, to fluoresce and possibly to explode. Observations of P/SW1 indicate that the outbursts develop in less than twenty-four hours. Various triggering mechanisms could activate a small area, such as an infalling boulder from space, as suggested by Martin Harwit at Cornell, or conceivably high-energy particles from a solar flare.

But how does the burst propagate over such a large area? I favor the idea that initially a small area of activity pushes out some larger clods. Some of these fall back on other areas that are nearly ready to burst out. When the surface is broken, these prepared areas become exposed and thus can carry on the chain reaction by pushing out more large clods that fall back on new areas. The spreading stops at the edge of the region of warmed subsurface complex ice. The total burst also stops in late afternoon (comet time) to resume next morning, until finally the layer of warmed complex ice has sublimated away. In polar regions turned toward the Sun, the burst may last several Earth days until the prepared regions are exhausted.

Paul Feldman and Michael A'Hearn propose that the huge night-day temperature changes on P/SW1 crack the surface by simple thermal expansion and thus initiate a burst. They believe, however, that there may be enough carbon dioxide, or possibly carbon monoxide, to provide the volatility needed for the outbursts. But, again, why does the process stop? The idea of "pockets" of, say, carbon dioxide on comets seems appealing at first scan. However, these "pockets" must be hundreds of meters in dimension and only centimeters thick, very strange kinds of "pockets," indeed.

Perhaps the complex ice is layered, the comet having grown from time to time in clouds of different compositions. Or perhaps small comets, formed of complex ice, fell into a larger comet and flattened out to produce various layers over the nucleus. Even though we still do not really understand how P/SW1 activates its great displays, the subsurface warming of complex ice does show great promise as a means of explaining the phenomenon. This ice could be widespread in the comet, and thus the comet would not need to have a highly unusual structure for the outbursts to recur time after time.

The puzzle of comet SW1 and other comets that are vigorous at large solar distances seems difficult enough to solve. It may be more easily explained, however, than the puzzle of comets splitting when isolated alone in space. In such cases, large pieces are mysteriously removed, gently, but sometimes, as in the case of comet West of 1976, with the sudden release of a great amount of dust and gas. We have seen that comet splitting is not likely to be the separation of two or more components of a double or multiple comet because a separation of that kind would not produce unusual activity. Also there is no evidence to suggest that comets spin up to throw off large pieces at the equator, unless possibly the nuclei were unusually elongated or flattened. Could complex ice possibly offer a solution to comet splitting?

Tempting is the idea that the outer layers of some comets may blow off with the accumulation of gas pressure under large areas because of internal warming. Complex ice might be slightly warmed well below the surface by conduction of solar radiation and eventually blow off large sections of the comet. Much more practical and theoretical research is needed to make this idea plausible. Internal radioactivity is another possible source of internal heating, especially for large comets, tens of kilometers in diameter. The heat would be carried outward by the most volatile gases, which might freeze again in the outer layers. It is difficult to believe, however, that the outer layers could be gas tight and could allow much pressure to build up over billions of years. Also, because radioactive heating falls off rapidly over such long periods of time, it is an unlikely cause of splitting in comets that have lived in the Öpik-Oort cloud.

At the same time, radioactive heating may explain why very old and originally very large comets, such as P/Encke, seem to behave as though water-ice were the only active ingredient. Calculations that Robert P. Stefanik and I made in 1965 show that the central regions of comets more than 8 kilometers in diameter would be sufficiently heated by radioactivity to cause a phase change in the cold complex ice. In the cores of larger comets, amorphous ice would have changed over to ordinary ice. The heating would have been so slow that explosions seem unlikely. In extremely large comets, 80 kilometers or more in diameter, radioactivity may have actually eliminated water-ice from the center, leaving a rocky core surrounded by water-ice. Such a comet, finally tamed into an asteroidal orbit, might be indistinguishable from an ordinary asteroid. Indeed, a few asteroids do move in orbits much like those of tame comets; their true origin has been a long-standing puzzle.

Radioactive heating may also explain why P/Encke ejects larger meteoric particles but no fine dust. The warming of its core (which is the part of the comet we now observe) may have caused the fine grains to stick together as large particles, or dust balls.

In summary, we have deduced that if comets form at very low temperatures—a few tens of degrees above absolute zero—the water they contain can freeze into exotic, amorphous forms I call complex ice. When warmed, complex ice gives off heat and possibly explodes, providing us with activity at much lower temperatures than would be needed to induce activity in comets made out of ordinary crystalline ice. Furthermore, laboratory tests show that comets exposed to cosmic rays in space for billions of years have the potential to develop an outer layer of mangled molecules that are very easily activated by the Sun at great solar distances, beyond Jupiter's orbit at more than 5 AU. This may account for the early brightening of comet Kohoutek and may also explain why we find many more wild comets coming in from the Öpik-Oort clouds than we find comets returning again. The loss of the outer, active layer on the first approach to the Sun may make the comet so much fainter on a subsequent return that our comet hunters do not discover it.

Greenberg's laboratory experiments indicate that, if we make comets by freezing out water and other compounds (of C, N, O, and H) in the presence of dim starlight at extremely low temperatures in deep space, they will contain complex ice and many highly complicated compounds. Complex ice may be expected to fluoresce and possibly explode when slightly warmed. At temperatures above about 153° K (−185° F), it reverts to ordinary, although highly contaminated, ice. Complex ice may be the source of activity in distant comets and may help to explain splitting comets.

We are now in position to talk seriously about how to make comets and picture something about the nature of comet factories—or are they nurseries?—deep in interstellar clouds, or around newborn stars such as our Sun in its infancy some 4.6 billion years ago. But before turning to this next phase of comet theory, let us mention briefly a theory of cometary origins that goes back to the seventeenth century. The enthusiastic observer and student of comets, Hevelius, suggested that comets may be thrown out by the great planets Jupiter and Saturn. Over the centuries, a few astronomers have favored the idea, especially after the Jupiter family of comets became evident from studies of orbits. In the twentieth century, the late Sergei K. Vsekhsvyatskii, a Russian astronomer, was a vigorous and lonely supporter of the planetary origin of comets. Even though the idea must be abandoned for obvious reasons, Vsekhsvyatskii's massive collection of the physical observations of comets will endure as an important standard reference for students of comets.

Chapter 21 How to Make Comets

Although our detective story can never really end, we now have enough clues to devise a basic recipe for making comets. The ingredients are various elements from interstellar clouds of gas and dust, specifically, elements and compounds that can freeze out at very low temperatures. Next, we must collect this dirty frost into sizable bodies, up to dozens (or hundreds?) of kilometers in diameter. The temperature in our deep-freeze comet factory or nursery must be carefully controlled. Our finished product will then be ready to be put into proper orbits about the Sun, which itself was formed from the same or similar interstellar clouds.

A more detailed recipe depends on theories of *where* and *when* the comet factory went into production. The outskirts of the infant Solar System when the Sun and planets were forming is the first choice of most astronomers today. This setting will require three variations in the recipe and three scenarios for placing the comets into the Öpik-Oort cloud gravitationally attached to the Solar System. We cannot ignore, and will briefly examine, the fourth possibility that comets were made elsewhere in different interstellar clouds, then subsequently captured by our Solar System. This idea introduces more complica-

Classical concept of the Laplace nebula and the development of the Solar System.

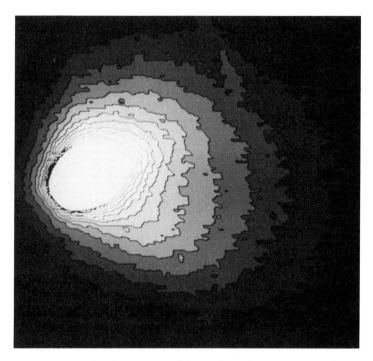

Curves of equal brightness around the nucleus of comet IRAS-Araki-Alcock on May 11, 1983. Width of picture: less than 2,000 kilometers at the comet. (Courtesy Rudolph E. Schild, Whipple Observatory.)

tions and uncertainties about the origin of comets.

The unanticipated 1983 comet IRAS-Araki-Alcock provided a vital clue to where comets are made when it came 4.6 million kilometers from the Earth. This clue is sulphur in the form of the rare laboratory double molecule, S_2. Its discoverers, Michael F. A'Hearn with David G. Schleicher at the University of Maryland and Paul D. Feldman at Johns Hopkins University, utilized a sophisticated Space Age tool: the satellite International Ultraviolet Explorer (IUE). The S_2 spectral lines in the near ultraviolet showed up, but only close to the nucleus. They had not been observed in other comets because sunlight destroys S_2 molecules in about ten minutes at 1 AU from the Sun. For a comet at 1 AU from the Sun, this amounts to a distance of about 300 kilometers from the nucleus. Only the near approach of comet IRAS-Araki-Alcock to

The International Ultraviolet Explorer Satellite. (Courtesy National Aeronautics and Space Administration.)

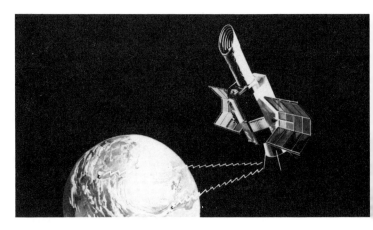

the Earth made this critical discovery possible, but its significance was not instantly apparent.

In their efforts to explain the presence of S_2 in the comet, the discoverers ran into difficulty. When a chemist cools hot sulphur, much larger molecules than S_2—such as S_8 and others—are usually produced. From the experience of chemists in trying to make S_2, A'Hearn and his colleagues concluded that the only way to put S_2 into a comet is to freeze out sulphur at extremely low temperatures in the presence of ultraviolet light or energetic particles. The temperature must not rise again until the comet comes close to the Sun and the molecule is blown off. An earlier warming will cause the S_2 molecule to combine with its neighbor molecules and disappear from the compound.

Note the importance of this requirement! It supports our assumptions about complex ice. It limits the temperature to very low values during the entire process of comet formation, indeed, until the surface is finally heated by the Sun. It thus encourages us to build comets from fine grains in the presence of radiation or energetic particles. These are precisely the conditions we expect in interstellar gas and dust clouds, and remind us of Greenberg's experiments and his concept of interstellar grains. The gas and dust clouds in the Milky Way may last for up to many millions of years as they swirl around near the plane of the rotating system. When turbulence compresses them enough so that their self-gravity can pull them together, they collapse to make clusters of new stars, perhaps solar systems.

In a cold, dense molecular cloud, a grain may be born as a microscopic particle of rocklike material. It may then grow a thick mantle of complex ice. New stars also may be forming nearby. Perhaps a new star may warm the grain as the star suddenly brightens. A brilliant nova or supernova could be close enough to heat up a huge "hole" in the dense cloud. The warming of the grain may sublimate away the complex-ice mantle, leaving a very thin layer of gunk on the rocky core. Greenberg visualizes a single grain passing through this sequence a number of times—gaining, losing, and regaining its mantle of complex ice. In each cycle, the rocky core of the grain develops a thicker layer of gunk. Greenberg suggests that, if comets do grow from such interstellar grains, the "missing carbon" that Delsemme fails to find in comet gases may actually reside in the yellow gunk on the cores of the grains. The unusual sulphur (S_2) molecule in comets would be one of the many compounds accumulated in the gunk.

To make a comet, we must collect these interstellar icy grains into kilometer-sized bodies without substantially heating the grains. Indeed, we need to use the grains as seeds to freeze out more ice and collect more rocky elements. The most obvious place and time to make comets is on the outskirts of the Solar System as it was forming. The idea of making the Solar System by the gravitational collapse of massive, low-density material in the Universe goes back to the eighteenth century—to the immortal philosopher Immanuel Kant and to Marquis Pierre-Simon de Laplace, noted French mathematical astronomer. Laplace provided the more detailed theory, his famous *Nebular Hypothesis*, in which he explained how many of the spiral nebulae could be "solar systems" in the making. That we now know these nebulae to be giant galaxies, each containing up to hundreds of billions of stars, cannot detract from our admiration for those great thinkers who, working with so little basic knowledge, were able to visualize appropriate processes to explain the origin of stars and solar systems.

Precisely how a section of an interstellar cloud col-

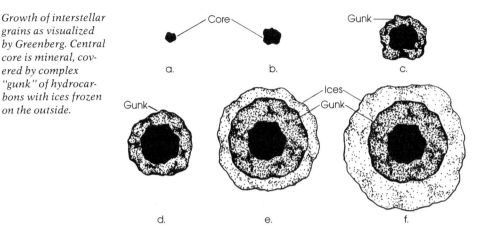

Growth of interstellar grains as visualized by Greenberg. Central core is mineral, covered by complex "gunk" of hydrocarbons with ices frozen on the outside.

lapses gravitationally into a star—a double or multiple star, or a solar system—is still a challenging theoretical problem. But astronomers have identified many of these stellar delivery rooms and nurseries in dusty clouds that are illuminated by massive young stars and by the novae. The Great Orion nebula is a fine example, being only about 1,600 light-years away. Undoubtedly, a collapse accelerates rotation in the original cloud, so that a new star is formed in a flattened, nebulous disk, much like that visualized by Laplace. The problem for the theoreticians is to slow down the rotation enough so that the system can hold together and actually make a single star. Any slight rotation in the original cloud is greatly magnified during the collapse. Hence the theoretician must find some way to dampen out the rotation or else to conclude that the formation of single stars like the Sun is extremely rare. Actually, most stars observed are double or multiple systems. Unfortunately, all stars except the Sun are so far away and any planets about them so faint in their reflected light that we have no direct knowledge at present of any planet belonging to any other star. A "brown dwarf" star discovered as a faint companion to the star Van Biesbroeck 8 is in no sense a planet. It is observed in its own light and is many times more massive than Jupiter.

An exciting discovery by the infrared-sensing IRAS satellite gives us some hope that planetary systems may not be too rare. In 1984, the IRAS scientists found that the fifth brightest star in the sky, Vega of the constellation Lyra, is surrounded by a cloud of warm grains. Vega is a relatively young star, less than a billion years old. We might consider it an adolescent star that escaped from or blew away its birth cocoon of interstellar gas and dust. It has kept or acquired, however, a cloud of particles that reaches out to about twice Pluto's distance from the Sun. The particles must be at least buckshot size or larger; otherwise, they would be lost by now. Whether planets have formed around Vega, whether they are now developing, or whether comets are prevalent, we can only speculate. Paul Weissman of the Jet Propulsion Laboratory has suggested that Vega's cloud may be maintained by grains released from a cloud of comets analogous to the Öpik-Oort cloud. A few other stars have since been found to be surrounded by such clouds of particles. When will we be able to prove that other planets exist in the universe? Perhaps soon, with the aid of instruments on the Space Telescope!

The entire Milky Way. (Composite photograph by the Lund Observatory, Sweden.)

Astronomers have yet to find an interstellar cloud in the actual process of collapse. Undoubtedly, the collapse phase takes place relatively quickly and occurs in great molecular clouds so dense that the dust hides the collapse

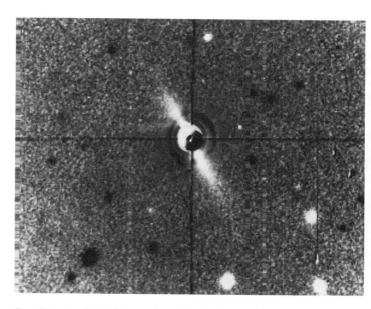

Beta Pictoris. Optical image from Earth of a possible planetary system in the making about another star. Beta Pictoris is 53 light-years away, so that the greatest radius of the disk is 400 AU, or about ten times Pluto's distance. (Courtesy Bradford A. Smith, who with Richard J. Terrile used extremely sophisticated techniques with the du Pont 2.5-meter reflector at the Las Campanas Observatory in Chile to obtain this computer-generated image. The IRAS satellite had discovered the infrared disk about the star. A central opaque disk occults the star at a diameter of 100 AU around it.)

from our view. But astronomers have found infant stars in the outer regions of great molecular clouds on the side toward us, where dust absorption is not too great. These infant stars are still embedded in dense cocoons of gas and dust left over from the collapsing cloud. For example, Eric E. Becklin at the University of Hawaii and Gerry F. Neugebauer at the California Institute of Technology have identified such an object in Orion. It appears as an infrared hot spot in the cloud core, and may consist of more than one newly born star. The surrounding cocoon results from the remnants of the collapse and is lighted by the early brightening of the star or stars. Light pressure and massive stellar gales heat and expand the surrounding gas while forcing the fine grains away from the center. In this way a baby star breaks free of its swaddling cocoon.

Fledgling stars, in a somewhat older age group, identified as T Tauri stars, were first recognized as young stars nearly three decades ago by George F. Herbig at the Lick Observatory in California. The T Tauri stars, named after the prototype in the constellation of Taurus, show bright spectral lines, vary in light, and blow off massive stellar winds.

Among the various interstellar clouds of the Milky Way, the giant molecular clouds (GMCs) are ideal birthplaces for both stars and comets. Astronomers have identified more than fifty interstellar molecules in these clouds, primarily by millimeter-wavelength radio detection. Some of the hydrocarbon molecules contain large numbers of atoms. Even ethyl alcohol (CH_3CH_2OH) has been detected.

A negative showing the nebulosity around a young star, FU Orionis. (Courtesy George H. Herbig, Lick Observatory.)

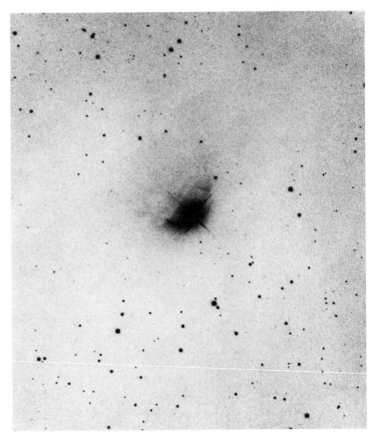

Many of the compounds, therefore, are already complex, even before being frozen onto cometary grains to form complex ice. The temperatures are extremely low in the GMCs, and their typical space densities are estimated to be 5,000 hydrogen atoms per cubic centimeter, which is many thousand times greater than the average in space. These clouds contain up to a million solar masses. The star-forming regions in them are much less massive, but are also much denser than is average in these great clouds.

When, according to present evidence and theory, the Sun and Solar System formed 4.6 billion years ago, almost certainly in a GMC, the cloud developed into a flattened nebulous disk, much like the one Laplace had envisioned. Opinions differ as to the original mass of the Laplace nebula, but most theoreticians believe it was greater than about 1.2 times the present solar mass, and probably not more than 2 times. There is also uncertainty as to the precise rate at which the Sun grew and developed, compared with the planets and asteroids. In any case, the outskirts of this nebulous disk provided an ideal environment for making comets. Theory shows that the temperature of the collapsing cloud remains extremely cold, except near the region of a solar embryo. Even after the embryonic Sun began to radiate appreciably, the gas and grains in the plane of the disk shielded the outer regions from much temperature rise.

Comets almost certainly must have formed in the same regions as the two great outer planets, Uranus and Neptune, because the average composition of each of these planets is very nearly the value we should expect if about a dozen Earth-masses of comets had been aggregated together to form each of them. Also the fact that many of their satellites are icy suggests cometlike material. We must now deal with a fundamental question. What about the comets we see today? Are they the leftover building blocks of these outer planets, later thrown out to the Öpik-Oort cloud by the giant planets acting as gravitational slingshots? Or were our present-day comets formed in the collapsing interstellar cloud or clouds, and were their original orbits much the same as they are today in the Öpik-Oort cloud? Or could our comets have been made in other GMCs and captured later?

For these three alternatives we can recommend three recipes: one for making comets in the outer planetary regions of the Laplace nebula, another for making them in dense cold GMCs, and a third for growing them as the Laplace nebula collapses.

We know best how to make comets in the outskirts of the idealized Laplace nebula, which is rotating and flattened, whereas the mechanism for making them elsewhere is, as we shall see, subject to many theoretical and factual uncertainties. In a rotating nebula, the system flattens out like a pancake, except for dying turbulent eddies from the infall of the outer interstellar cloud. Remember that some 98 percent of the nebula is hydrogen and helium, and is always gaseous. Our cometary mix is less than 2 percent of the total, the maximum amount we may hope to freeze out.

A theory for making planets near the forming Sun in a flattened rotating nebula was developed by Peter Goldreich and William R. Ward at the California Institute of Technology in 1973. Their theory holds for comet-building at the outskirts of the Laplace nebula if one makes the proper assumptions, which may or may not have been applicable in our actual Solar System. Because great compression occurs in the collapse, tiny grains of interstellar dust grow much more rapidly than they could in the rarified interstellar clouds. When the grains have grown to birdshot dimensions, their motion ceases to be controlled completely by the gas. Each pellet moves more or less

Cross section of one-half of the Laplace nebula showing the thin layer of grains settling out from the gas above and below (in the Goldreich and Ward concept).

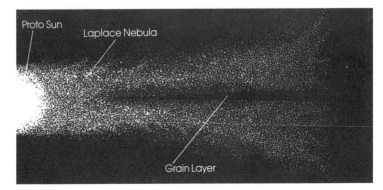

Proto Sun

Laplace Nebula

Grain Layer

independently of the gas, and its orbit is controlled mainly by the central gravity of the growing Sun and the inner nebula. At this stage, the orbit of a pellet may be somewhat eccentric and inclined a few degrees to the average plane of the gaseous nebula. Each pellet in each revolution around the center is moving up and down, and in and out, with respect to the gas. The gas, in turn, is now rotating about the growing Sun. The drag or resistance of the gas slowly dampens the relative up-and-down, in-and-out motions of the pellets, causing them to settle rather rapidly to the average plane of the nebula and to move in nearly circular orbits.

Once the comet pellets are tightly concentrated into a very thin disk in the plane of the nebula, according to the theory of Goldreich and Ward, their mutual gravitation breaks up the disk so that the pellets initially clump together into small cometesimals. These in turn aggregate more slowly into larger and larger bodies. In the outskirts of the nebula, they can be considered comets. If this happens where the nebula is massive enough to grow the great planets Uranus and Neptune (and Pluto), then a large number of larger bodies—possibly the size of the Moon or even the Earth—might successively collide and coalesce into these planets. Perhaps Uranus acquired the 98-degree-tilt of its spin axis to its orbital plane by a late hit from one of these overgrown comets.

In this Goldreich-Ward scenario, the leftover building blocks of Uranus and Neptune—the comets—were successively disturbed by close encounters with huge comets and finally by the finished Sun and planets. This activity threw them helter-skelter all over the Solar System and many out to infinity. Jupiter and Saturn may have captured a few Earth-masses of these comets each and helped kick some of the remainder about. A huge number died normal deaths in orbits near the Sun, some falling into it, while a few Earth-masses of them, perhaps a hundred billion or more comets, ended up with elongated orbits in the Öpik-Oort cloud. All of this may have taken place over a period of 10 to 100 million years. Most astronomers probably would favor a scenario similar to this one for comet formation.

A. G. W. Cameron of the Harvard-Smithsonian Center for Astrophysics, suggests an alternative scenario that still uses the same comet-making recipe in a different region in space. He starts with a very massive solar nebula having about twice the mass of the Sun, and makes comets well beyond the areas of Uranus and Neptune, at 1,000 AU or more from the center of the system. The Sun and inner planets form rapidly, perhaps in a hundred thousand years, at which time the young Sun's violent activity quickly blows out the remaining gas and dust of the solar nebula, reducing the central gravitating mass by about half. The planets, most of which have formed by that time, expand their orbits about twice in size because of the reduced central mass, but otherwise are not greatly affected by the loss. The planets make many revolutions while the nebula is being blown away, spiraling out gently. Comets, on the other hand, with orbital periods of several tens of thousands of years or more, rather "suddenly" find themselves moving about a central mass that has been reduced by half. They therefore fly out into huge orbits, creating the great Öpik-Oort cloud. Then, as Öpik first showed, passing stars randomize their orbits and, on the average, increase their perihelion distances. The stars kick some of them in for us to see, by Oort's process, while kicking many of the others out into the Milky Way. Only time will tell the fate of Cameron's theory in the scientific arena.

But we have no proof that the comets we see actually grew in the Laplace nebula of the forming Solar System. The comets of the Öpik-Oort cloud might have been made in the swirling and infalling nebulous matter. No one has investigated this possibility thoroughly, because it involves so many uncertainties. We have no knowledge of, or even a basis for, estimating how large or dense the collapsing cloud may have been or the number of sibling stars in the Sun's nursery. In any case, Brian Flannery and Max Krook at Harvard have shown how comets might form in very cold and dense GMCs. And Jack G. Hills at the Los Alamos National Laboratory has developed a theory in which the comets of the Öpik-Oort cloud were formed during the infall of the interstellar clouds as they collapsed to make the Solar System.

The mutual shadowing of two particles from diffuse interstellar radiation produces a net attractive force between them.

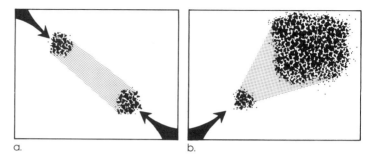

a.   b.

Let us first consider the Flannery-Krook recipe for comet making. In a dense cold GMC, the grains will grow their relatively thick coats of complex ice. In deep space, where no special force acts on the icy dust grains, they will move along with the gas. They will collide with each other so seldom that they will not be able to grow much by accretion. We must find some force that will act selectively on the grains, not the gas, and bring them closer together. Light pressure is a likely candidate. Lyman Spitzer, then at Yale, noted in the 1940s that even a dust grain will cast a shadow. Suppose that two grains are near each other in deep space, where diffuse starlight comes from all directions. Each grain stands in the shadow of the other. Each grain thus experiences more light pressure on its side away from the other grain than on its near side. As a result, the grains are pressed toward each other according to the inverse square of the distance. A larger grain or a clump of grains will attract single grains more strongly, the attraction being proportional to the shadow area. After enough grains have moved through the gas into compact groups, their mutual gravity enables the light pressure to cause them to grow into cometary masses, according to the Flannery-Krook theory.

In the Hills scenario, after the Sun began to shine vigorously in its stellar nursery—where other and much brighter, large, baby stars were also shining—the Sun's light pressure made comets grow. Under the pressure of light from the radiating Sun, the interstellar grains shadowed each other, as we have already described, so that they bunched together, this time in huge, thin sheets roughly centered about the protosun and the Laplace nebula. When they were sufficiently concentrated, their self-gravity drew them into large, dirty snowballs, or comets. They were moving in great orbits, many hyperbolic as far as the Sun was concerned, but enough ended in closed orbits about the Sun to populate the Öpik-Oort cloud. I have worked on this type of theory, but never published the results because I did not feel confident that the processes would really work out. Hills and his colleague M. T. Sandford II in 1983 delved into the theory and calculations in a thorough fashion, and demonstrated that it has promise. But the theory rests on so many uncertain assumptions that its validity is questionable. No consensus of informed opinion about it is yet available.

The theory that the comets we see were made in interstellar clouds, and then captured into the Öpik-Oort cloud about the Solar System, seems fanciful because it lacks a sound footing either in theory or in observation. We know nothing about the actual existence of comets in interstellar clouds, because such comets would be completely unobservable from Earth unless they happened to pass very near to the Sun. As discussed earlier with respect to the Öpik-Oort cloud, we doubt that any observed comets are moving in hyperbolic orbits about the Sun. Nonetheless, the chance of our seeing such comets is so extremely small that interstellar clouds could contain huge numbers of comets, quite beyond our ken. Indeed, an appreciable fraction of the huge mass of interstellar clouds may actually be comets.

The possibility that interstellar space may be teeming with comets has led William Napier and Victor Clube of the Royal Observatory at Edinburgh, Scotland, to advocate that the Öpik-Oort cloud of comets did not originate with the Solar System. They maintain that comets are captured every few hundred million years, as the Sun passes through denser interstellar clouds in its 250-million-year circuit of the Milky Way. They calculate that the Öpik-Oort cloud could not survive these passages and must be replenished. Most astronomers believe that the Sun is too inept (theoretically) to capture such comets, if they indeed exist, so the theory has as yet gained little general acceptance.

If we could recover a comet sample via a space probe and measure its age by radioactive isotopes, we might settle the question once and for all. An age younger than 4.5 billion years would indicate that it must have been made later than the Solar System and in all probability was captured from an interstellar cloud. A greater age would leave the question partly unanswered, because the comet might have been made along with the Solar System but from older interstellar dust, or it might have been captured. An age equal to 4.6 billion years would indicate that comets formed with the Solar System, either in the outskirts or perhaps during the collapse, as Hill proposed. Such an age would strongly support the early formation and longevity of the Öpik-Oort cloud of comets.

In this chapter, we have looked at four scenarios for making comets and putting them into the Öpik-Oort cloud. The first and most broadly accepted (or should I say tolerated?) theory is that the comets formed in the region of the outer planets by the Goldreich-Ward concentration method, and were then kicked into the Öpik-Oort cloud by the great planets, notably Uranus and Neptune. Comets were almost certainly made in this fashion because Uranus, Neptune, and several of their satellites show the proper densities and probable compositions to have been made out of comets. A nagging question remains, however: Are the comets that we see the leftover building blocks of these bodies, or were they formed farther out from the new Sun and planetary system?

The second theory is Cameron's modification of the first, that the comets formed well outside the planetary region, but within the Laplace nebula. The argument here is that comets were thrown into the Öpik-Oort cloud by the sudden ejection of a very massive solar nebula, as the Sun developed into a bright new star. This theory will require a great deal of substantiation if it is to be accepted by most astronomers working in the field.

The third theory, carefully investigated by Hills and Sandford, suggests that comets were made in the interstellar clouds as they were collapsing to form the Laplace nebula and the Solar System. Light pressure from the infant Sun and from its stellar siblings is the prime mover in concentrating the interstellar dust. A distinct merit of this theory is that it places many comets immediately in the Öpik-Oort cloud. Whether the theory is acceptable remains uncertain because of its newness and basic uncertainties about conditions in and around the Laplace-type nebula. But certainly the theory shows enormous promise.

The fourth theory involves the capture of comets from passing interstellar clouds, in which the supposedly abundant comets had been made by processes akin to those described by Flannery and Krook, or by Hills and Sandford. In this theory, supported by Napier and Clube, the comets may have been formed at any time before the present, not just at the time of the formation of the Solar System, some 4.6 billion years ago. However, the probability of capture required in this fourth theory appears to be very small, if, indeed, interstellar clouds are teeming with comets. More detailed study is required to make the capture theory acceptable.

Although we cannot be certain where or precisely how our comets were made, the clues that we have collected give us a clear qualitative picture of their origin. Comets grew from cold, icy grains. Most astronomers believe that they aggregated in the general neighborhood of the Sun when it was young. Certainly, many comets have fallen onto the Earth. Have they in any way contributed to our most important possession—life?

Chapter 22 **Comets and Life on Earth**

Although there are few substantive facts relating comets to the presence of life on Earth, we will briefly speculate about extraterrestrial origins of life and the possible role of comets in making Earth a hospitable environment for the development of life. Historically, most cultures have provided a simple answer for the origin of life, the details depending upon the culture: An omnipotent supernatural being or some unknown life-force initiated life on Earth. Many scientists, especially biologists, whether or not they have accepted the dogma of their religion, have proposed a number of hypotheses as to the mechanism. However, the origin of life remains a scientific enigma that still baffles our best minds, even though they are now armed with the latest tools of modern science.

Some people believe that life began far from the Solar System somewhere in the depths of space. A staunch supporter of this idea was Svante A. Arrhenius, the great Swedish chemist and Nobel Prize winner. In his book *Worlds in the Making* (1908), he suggested that living spores traveled through the depths of space to initiate life on Earth. He postulated that the then recently discovered pressure of light spread these spores around the Universe.

In some more modern space travel scenarios, extraterrestrials (ETs) have *intentionally* seeded the Earth with life in various ways—by colonizing, by sending capsules to Earth loaded with appropriate life forms, or even by dispersing large numbers of capsules throughout the Galaxy in the hope that some would find landing sites on congenial planets such as the Earth. In alternative proposals, the ETs have done so *unintentionally*—by means of derelict spacecaft, or by leaving their garbage in space near the Earth or at landing sites on Earth. In a more dismal scenario, another planet developed intelligent life but was blown up by its not-so-intelligent inhabitants. Some of the surviving life forms then traveled through space to the Earth. Or, the planet suffered an accidental catastrophe such as a collision, and thereby spread life forms through space.

Among these entertaining speculations we can, at last, rule out ones that assume longevity for spores, bacteria, viruses, or other living organisms in space. In 1984, J. Mayo Greenberg subjected the bacterium that is most resistant to ultraviolet light, *B. Bacillus subtilis*, to laboratory tests. Under exposure to ultraviolet light equivalent to only a thousand years in space, the bacillus was completely destroyed! The addition of other known hazards in space such as X rays, cosmic rays, and solar-wind ions, would surely have hastened the process. Altogether, near-Earth space is a death trap for any type of living organism we know.

We could well end this section of the chapter here, were it not for a proposal made in the late 1970s by Sir Fred Hoyle of Cambridge and Chandra Wickramasinghe of Cardiff University, namely that life on Earth originated in comets. Their widely publicized theory states that "comets carry, amplify, and dispense life throughout the Universe," including plagues, influenza, and all types of obnoxious viruses. Their comets are the same dirty snowballs that we have discussed. Because, chemically, comets are highly heterogeneous—so their theory argues—they should contain all the major and minor elements required for the production and sustenance of life. To make comets congenial hosts for the genesis of life, Hoyle and Wickramasinghe assume that radioactivity in the cores of comets produces (or produced) warm, central pools of water. Thus our first basic question concerns the likelihood that such warm pools can actually exists in comets. If this is likely, the next question is: Can they last long enough for life to develop from the basic elements? As mentioned earlier, in 1966, Stefanik and I calculated that the natural radioactivity to be expected in comets could indeed produce enough heat to melt water at the cores of comets much greater than 30 kilometers in diameter.

Any warm pools must be insulated and sealed inside a comet by a shield of several kilometers of dirty ice to protect life forms against the cold and vacuum of space. Heat conduction is slow in icy rubble, the heat being carried

Harold C. Urey.

mainly by the evaporation of volatile ices, which freeze again in the outer layers. The presence of a liquid increases the rate of heat flow and loss by a huge factor. Furthermore, no one can calculate confidently the factors of cracking, expansion, or collapse—that is, the general stability of the outer protective layers. Comets do split without apparent cause. Is it not pure fantasy to propose that warm pools of water can exist for many hundreds of millions of years in any but the largest comets we know of, if, indeed, in any at all? Futhermore, typical comets are much too small to support warm, central cores.

Yet, even if we accept the possibility of warm pools in comets that are the conveyors of life forms, we cannot ignore another factor that mitigates against life originating in them: the lack of an energy source except a mild temperature to produce the miracle of life. The famous experiments performed in 1953 by Stanley L. Miller and Nobel Prize winner Harold C. Urey showed that simple molecules of hydrogen, methane, ammonia, and water combine into amino acids—the precursor building blocks for life itself—when subjected to electrical discharges similar to lightning. In comets, amino acids may have been formed on the interstellar grains that made up the comets. There can be no light and no additional energy source deep inside comets, however, except for a small amount of radioactivity—which is clearly inimical to, rather than supportive of, life. The process is still baffling under ideal circumstances. How it could occur in the blackness of a comet core strains one's credulity.

On Earth, we certainly had warm pools of primordial "soup" for hundreds of millions of years. To me, primordial Earth presented a far more suitable environment for the origin of life than the highly speculative warm pools in huge comets. Furthermore, the warm pools on Earth were subjected constantly to light and lightning—energy sources for myriads of complex chemical reactions extending over eons of time. What better incubator for life?

With regard to the dispersal of life through the Universe, I agree that sizable bodies, such as comets, are required. Any life forms must be carried in bodies on the order of meters in dimension, or, preferably, very much larger. Otherwise, ultraviolet light, cosmic rays and high-energy particles would destroy life forms during the huge periods of time required to carry them through astronomical distances. Furthermore, the life-carrying bodies would have to deposit the life forms safely on the surfaces of planets with appreciable atmospheres, that is, planets somewhat like the Earth. These impactory encounters, at speeds of several kilometers per second, produce fireballs that vaporize a considerable thickness of the carrier. Again, sizable carriers would be needed to protect the precious life forms from being boiled away in the process of landing on the planets. Comets probably fulfill these size requirements, even though the circumstances of entry are far from ideal.

The Hoyle-Wickramasinghe proposal rests on three assumptions, each of which is highly improbable and all of which must be correct: (a) that warm pools continued to exist in the cores of comets for hundreds of millions of years, (b) that life originated in such black quiescent pools devoid of energizing radiation, and (c) that the postulated

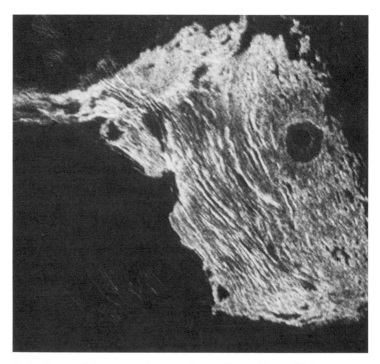

Radar image of the mountainous region on Venus, Maxwell Montes, whose circular feature (crater?) is about 100 kilometers in diameter. The folded mountain chains may be analogous to the Appalachians on Earth. The mountain rises to 11 kilometers above the mean surface of Venus. (Courtesy D. B. Campbell, Arecibo Observatory, Puerto Rico.)

ВЕНЕРА-14   ОБРАБОТКА ИППИ АН СССР И ЦДКС

Soviet Venera-14 image of the surface of Venus. The landing site is characterized by flat-surfaced rocks, perhaps indicative of intense weathering from the dense venusian atmosphere. (Photo courtesy the Bernadsky Institute, USSR.)

life forms were successfully deposited on Earth. Enough said!

Let us turn to the finite possibility that comets may have helped make the Earth suitable for the development of life. Our best evidence indicates that the Earth accumulated in a sequence of processes, beginning with dust that grew in the primitive solar nebula. After the dust reached birdshot size, it settled to the plane of this rotating, flattened nebula, as described in the theory proposed by Goldreich and Ward. Self-gravity collected the birdshot-sized particles into larger aggregates, which then kept on colliding to form larger masses; finally, they became Earth and the three other terrestrial planets, Mercury, Venus, and Mars. In this scenario, the Earth once had a primitive atmosphere collected from the nebula. The great bombardment that completed the building of the Earth heated the surface layers, while radioactivity heated the interior. Alternatively, the Earth may once have been the hot core of a larger gaseous protoplanet. In either case, life could not have developed so early in Earth history. The presence of only negligible amounts of the heavy noble gases such as xenon and krypton in today's atmosphere, proves (in either scenario) that any primordial atmosphere has been lost. Otherwise these noble gases should be much more abundant.

Today we find a thin veneer of water on Earth—its oceans—averaging about 2.5 kilometers deep for the entire Earth. On Mars, we see a trace of water in the form of snow and possibly a considerable amount in the form of permafrost. On Venus, almost a twin-sister planet to Earth, we find a huge carbon dioxide atmosphere with only a trace of water. In fact, the amount of carbon in the atmosphere of Venus is almost equal to the carbon now locked up in limestone rocks on Earth. The Earth may have gained a similar secondary atmosphere with about the same amount of carbon (carbon dioxide or methane).

After the Earth's secondary atmosphere formed, however, it had an excess of water, which combined with the atmospheric carbon to make a nearly kilometer-deep layer of carbonaceous rocks. When life started up on Earth, it removed almost all of the remaining carbon from the atmosphere, leaving oxygen and a small equi-

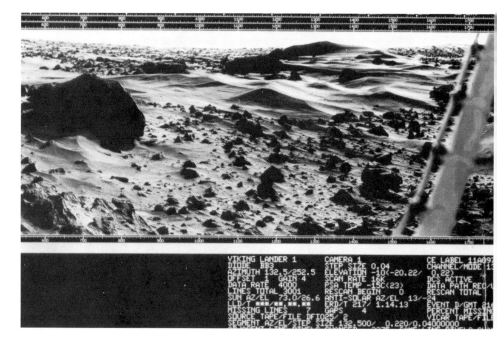

Mars terrain from Viking Lander 1. (Courtesy National Aeronautics and Space Administration.)

librium quantity of carbon dioxide. The lightweight hydrogen left over from the water-carbon reactions disappeared into space from the high atmosphere. Apparently Venus either gained too little water in its secondary atmosphere for carbonaceous rocks to form and deplete the carbon dioxide in its atmosphere, or else its suface temperature was always too high. In any case, Venus is the prime example of the greenhouse effect; carbon dioxide, being opaque to the heat radiated from the surface, maintains the surface temperature on Venus at 460° C (680° F).

We can only guess at the amount of water that Mars may once have held in its atmosphere. Because of its low gravity, it has certainly lost huge quantities of both the hydrogen and oxygen in water, dissociated by solar radiation and the solar wind in the high atmosphere. The surface rocks appear to be rusty as they consist of oxidized iron and other heavy elements, but the depth of these rocks and the quantity of permafrost on Mars awaits measurement by future space probes.

Our foremost question now becomes: Just how did the

Earth obtain its secondary atmosphere and the veneer of water over its surface? Similar processes must have taken place on Venus and Mars. Most geologists hold that seepage of trapped gases from the interior of the Earth was adequate to produce the present atmosphere and hydrosphere. Astronomer Tom Gold of Cornell University holds that the process is not yet complete, and that primitive methane (CH_4), from which most of the oil deposits have been made, is still seeping up slowly through the crust of the Earth.

These theories are not easy to evaluate quantitatively

The Earth as seen during the return of the Apollo 16 mission to the Moon. The western part of North America and the Baja peninsula are visible in this hand-held photograph taken in 1972. (Photo by NASA, courtesy Planetary Image Facility, National Air and Space Museum.)

MISSING—REWARD for location of missing Martian. (Flying Martian Cyclops by Donald H. Menzel.)

and definitively. So there remains the possibility that comets actually contributed to the life-giving elements on Earth, that some or many of the atoms in our bodies have come from comets. The comets that probably made Uranus and Neptune must have left behind a huge surplus to be kicked (gravitationally) toward the inner Solar System, outward to the Öpik-Oort cloud, and beyond to infinity. Perhaps hundreds of Earth-masses were involved in this wasteful process. Comets may well have contributed significantly to the water and carbon budgets of Earth, Mars, and even Venus. Eventually, the science of comparative planetology, coupled with space missions, may tell us whether we ourselves are made of comets, either in part or in entirety. In the meantime, the possibility provides an incentive for intensive, continuing exploration of the Solar System, which should lead to a more precise understanding of how the planets, comets, and the Solar System became what they are today.

Chapter 23　Comets May Be Dangerous to Your Health

This chapter title does not refer to old superstitions or to any danger that comets may bring viruses or poisonous gases to the Earth. It refers to the worldwide consequences of such a comet impact. Once comets were recognized as bodies moving through the Solar System, they became possible hazards in a real physical sense. Fear mongers often gained notoriety by embellishing the idea. Perhaps the first man to calculate the probability of collision with a comet was the astute and versatile French physicist Dominique Arago. In 1833 Arago published a book containing his conclusion that a comet in an average orbit, with perihelion inside the Earth's orbit, had "one woeful chance" in 281,000,000 of colliding with the Earth during one passage about the Sun. Some of his chapter titles indicate his concern about possible comet hazards: "Was the Deluge Caused by a Comet?" and "Has Siberia Ever Experienced a Sudden Change of Climate by the Influence of a Comet?" Note his interest in the possibility that comets may affect the weather. Other scientists have shown similar concerns. On June 30, 1861, the Earth may have passed through the tail of the great comet of that year. The well-known meteorologist E. J. Lowe noted that "the sky had a yellow, auroral glare-like look; and the Sun, though shining, gave but a feeble light." At 7:45 p.m., the comet was plainly visible despite sunlight, but had a much more hazy appearance than on subsequent evenings. Only after making these observations did he become aware of the possible Earth-tail encounter.

Although these possible effects of comets on the Earth may be coincidental and alone cannot be given serious consideration, we have direct evidence for a devastating impact on the Earth 65 million years ago, believed by a number of scientists to have eliminated the dinosaurs. Whether or not the dinosaurs could have survived the changing climate on the Earth if such an impact had not occurred is a moot question, but of considerable interest. The continuance of the dinosaurs might well have

A satisfied astronomer: "Yes, my dear friend, I have discovered a comet, and according to my calculations, I am in position to hope that it will collide with our Earth in forty-five days."

restrained or prevented the development of mammals and hence retarded, or even forestalled, the evolution of the human race. What actually happened? And what will happen again sometime in the future?

Paleontologists have long agreed that the fossil record clearly shows a massive extinction of a wide variety of living species, marking the end of the *Cretaceous* period, in which the dinosaus had been the dominant animal species. The extinction ushered in the *Tertiary* period, during which the mammals slowly gained supremacy. Modern isotopic-dating technology now places this worldwide event at 65 million years in the past. We need not dwell on the dozens of explanations that have been suggested for the extinctions. Suffice it to say that one of the theories, first suggested by Harold C. Urey two decades ago, and elaborated on by others later, attributes the catastrophe to the impact of an asteroid or a comet.

To determine the cause of the catastrophe, particularly to distinguish between an impact or possible heating by a nearby supernova, Luis W. Alvarez, his son Walter, Frank Asaro, and Helen V. Michel of the University of California at Berkeley, investigated the matter. They analyzed the deep-sea limestones at the Cretaceous-Tertiary (K/T) boundary in Italy, Denmark, and New Zealand. In 1980 they found a marked excess (20 to 160 times) of iridium at the bottom of the K/T stratum, compared with layers above and below. The K/T boundary was marked by a thin layer of clay about 2 centimeters deep, above which the excess iridium fell off rapidly over 7 centimeters. Iridium, along with some other heavy elements such as osmium and platinum, is extremely rare in Earth rocks. In meteorites it is also rare, but many times more abundant than on Earth, because chemically iridium tends to go with iron. Most of the Earth's iron has sunk to its inner core, the iridium disappearing with it, leaving crustal rocks greatly depleted in iridium in comparison with its average abundance throughout the Earth or in the meteorites.

Alvarez and his group calculated that an asteroid some 10 kilometers in diameter must have plunged into the Earth 65 million years ago. They based their estimates on the amount of unusual clay in the boundary layer and the amount of iridium to be expected from stony meteorites. What is the likelihood of such an event? The astronomical

The great Barringer meteor crater in Arizona. (Photograph by the late Moreau Barringer.)

expectations of an asteroid 10 kilometers in diameter striking the Earth is about once in 30 to 100 million years—a crude estimate that adds some support to the idea. Furthermore, the impact of such a stony body would be catasrophic. It would make a crater more than 150 kilometers in diameter and distribute a huge mass of dust into the stratosphere.

The dust from such an asteroid impact, according to Alvarez and his collaborators, would amount to about a thousand times that produced by the great volcanic explosion of Krakatoa island, between Java and Sumatra, in 1883. About 4 cubic kilometers of the Krakatoa dust was blown into the atmosphere and some of it remained in the stratosphere for perhaps two years. Three months after the eruption, the dust of sulphuric acid produced such brilliant red sunsets that New York City, Poughkeepsie, New York, and New Haven, Connecticut, called out fire engines to extinguish the apparent fires in their western suburbs! No conspicuous worldwide weather changes can be attributed to Krakatoa, but the average temperature may have dropped slightly. In contrast, the Indonesian Tambora volcanic explosion of 1815, even greater than Krakatoa's, is often blamed for the weather of 1816, which became known as the year without a summer.

The biota extinctions at the end of the Cretaceous period are attributed by Alvarez and his colleagues to the worldwide darkness produced by the dust. This thick layer of dust, they reasoned, would have caused the average temperature to drop by several degrees and would have prevented photosynthesis—the process essential to plant life both on the land and in the seas. Thus the food chain, from small to larger organisms, was blocked, with the result that all land animals greater than 25 kilograms in weight were killed and more than half the species of living organisms, both on land and in the sea, were destroyed.

Although most scientists agree that the calculated amount of dust and darkness would wreak havoc to life on Earth, they question whether it would be adequate to produce the overwhelming K/T disaster. For example, James B. Pollack and associates at the NASA Ames Research Center calculate that the total darkness and the freezing surface temperatures would last for less than a year. This finding only partly supports the conclusion of Alvarez and his colleagues. Other worldwide effects from the impact are to be expected. The fireball would produce a tremendous quantity of nitrogen oxides that would be converted to acids. Thus, acid rain would penetrate the upper 30 to 100 meters of the oceans and lakes. This acid would be particularly lethal to the calcareous organisms made of calcium carbonate or chalk. Noted examples are the coccolithophores, which are minute, floating, spheroidal, single-celled algae. These shield-like calcium carbonate structures built the famous white cliffs of Dover, England, before the K/T event. Since then, the coccolithophores have never recovered their important status in tropical oceans.

As a further hazard, carbon dioxide produced by the crater explosion might have contributed a hot period, following the periods of darkness and acid rain immediately after the impact. This carbon dioxide could have heated the atmosphere—the greenhouse effect—by blocking the heat radiation from the Earth to space. Other worldwide meteorological changes would surely have further damaged life on Earth. Note in particular the effects that would have been produced by such a gigantic impact crater. The resulting volcano could have added an enormous amount of steam and volcanic dust to the atmosphere, and it possibly erupted sporadically for centuries after being formed.

The evidence presented by Alvarez and his associates excited considerable interest and prompted many follow-up studies. By 1984, the iridium anomaly had been identified at some fifty sites around the globe, and the view that the K/T event was indeed an *abrupt* extinction caused by the impact of a large extraterrestrial body became strengthened. Nothing yet found in the deposits produced by the impact, however, has made it possible to determine whether the impacting body was an asteroid or a comet. The asteroid identification should be preferred, in my opinion. The study of asteroids whose orbits cross the Earth's orbit shows—according to Eugene M. Shoemaker of the U.S. Geological Survey and George W. Wetherill of the Carnegie Institute of Washington—that an asteroid 10

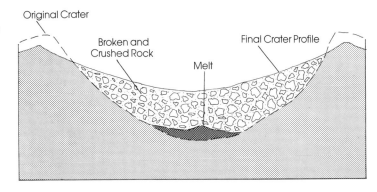

Schematic cross section of a high-velocity impact crater.

Original Crater

Broken and
Crushed Rock

Final Crater Profile

Melt

kilometers in diameter should strike the Earth about once in 100 million years. This rate is not inconsistent with the number of impact craters identified on Earth so far. Various estimates of impact rates for comets, as we see them today, lead to the conclusion that only faint and small comets, more or less a kilometer in diameter, should strike the Earth with this frequency.

Worldwide depth soundings have turned up no evidence of this or other great impact craters on the ocean bottoms. The reason is fairly well understood, and scarcely involves the shielding effect of the water. A 10-kilometer asteroid would make an instantaneous crater with a depth a number of times its diameter, a depth only slightly reduced by 2 or 3 kilometers of water. On land, most of the crater is filled in by fallback and slumping of the locally disturbed crystal rocks. The crust generally extends to much greater depths under the continents than under the oceans. The ocean crust is relatively thin, on the order of only 5 to 30 kilometers. Hence, such a crater would be largely filled by the subcrustal material, which is less viscous than the crust. I had the temerity to suggest that the great impact might have occurred on a very shallow and thin oceanic ridge. The crater, according to my scenario, then tapped the hot underlying magma to erupt into a volcano 200 kilometers in diameter that spewed out Iceland, the only island on an oceanic ridge. My suggestion has met with no scientific support, especially because Iceland appears to have formed a few million years after the K/T event.

What effects would we expect if the impact had occurred on a continental shelf? Probably the sediments would have filled the remaining shallow crater. More important, the ocean crust is relatively short-lived because of continental drift. Over periods of tens to hundreds of millions of years, an appreciable amount of ocean crust is subducted under continents and lost to observation. In view of the number of impact craters identified on land, a few of the more recent ones should be discoverable on the sea bottoms. Perhaps new sounding techniques will be developed to identify such craters, even though they may be filled to their limits. The K/T impact crater should have left enough geological effects to be found eventually, but it may already lie under a continent.

The broad acceptance of the K/T catastrophic event, with its biological consequences, has spurred a number of scientists to consider that impacts of asteroids and comets may have caused many, if not all, of the major extinctions identified by the paleontologists. These speculations are fascinating because of their implicatons with regard to biological evolution induced by the environment in the past and with regard to the possibility of extinctions in

Composite view of Phobos, inner satellite of Mars. Diameter about 23 kilometers. (Courtesy National Aeronautics and Space Administration.)

the future, including that of the human race. Even though many years may be needed to establish the scientifically valid aspects of these speculations, a brief recounting of the current discussions on the subject seems pertinent.

Craig B. Hatfied and Mark J. Camp of the University of Toledo suggested in 1970 that the great extinctions recur periodically and may be correlated with the oscillations of the sun above and below the plane of the Milky Way. The chance of the Solar System passing through a molecular cloud is greater near the plane because the clouds are more concentrated there; but this passing could happen even at the extreme limits, about 300 light-years above or below the plane. At present the Solar System is near this plane, with an uncertainty of about 2 million years, and is moving "upward" at about 6 kilometers per second. We are in a "hole" between interstellar clouds, where there is a minimum of gas and dust. Because the period of oscillation is roughly 67 million years, crossings occur at intervals of some 33 million years; these are close to the 32-million-year intervals at which extinctions appear to occur, as estimated in 1977 by A. G. Fischer and M. A. Arthur of Tulsa. Since then a number of scientists have addressed the subject. In 1983, David M. Raup and J. John Sepkoski at the University of Chicago placed the period at 30 million years after looking over the nine larg-

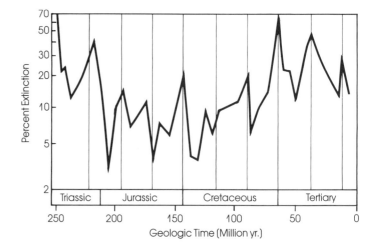

The 26-million-year repetition of biological extinctions, as plotted by Raup and Sepkoski. (Proceedings National Academy of Sciences 81, 1984.)

est extinctions over the past 250 million years. M. R. Rampino and R. B. Stothers at NASA's Goddard Institute of Space Studies, in December 1984, found an average periodicity of 33 ±3 million years for five different geological phenomena.

Napier and Clube, as noted earlier, calculated that the Öpik-Oort cloud could not survive passages through the great interstellar molecular clouds. In 1979, they suggested that comets are captured from these clouds and that rains of comets on the Earth produced the extinctions. They further suggested that both comet rains and extinctions are correlated with the dates of large impact craters on the Earth during the past 350 million years. Rampino and Sothers Stothers stress the near coincidence between extinctions and Milky Way crossings, and favor the idea that the terrestrial phenomena are modulated by rains of comets.

Various mechanisms other than rains of comets have been suggested whereby interstellar clouds might cause extinctions. The cloud material might penetrate the inner Solar System and raise the Sun's radiation by accretion, besides polluting the Earth's atmosphere and affecting the Sun's light and the solar wind reaching the Earth. There are so many possible effects from passages through interstellar clouds that the subject remains open to much more research—both theoretical and observational—in subtle geological analysis of the strata dividing the different eras. The occurrence of several ice ages in Earth history, for example, still demands more specific verifiable explanations.

All in all, a cometary and galactic cause for extinctions might not have many proponents, were it not for the record of great impact craters on Earth. Richard A. F. Grieve at Brown University has collated data for sixty-two great impact craters up to 600 million years in age. Geological and mineralogical signs of impact explosions identify the craters. Crushed rocks underlie their centers, and upturned strata at their walls provide important evidence. Often sections of the rocks show the signs of shock, where the grain structure is shifted. Impact shatter cones are sometimes encountered. Impact craters are dated by

Bright meteor that split. (Smithsonian Astrophysical Observatory.)

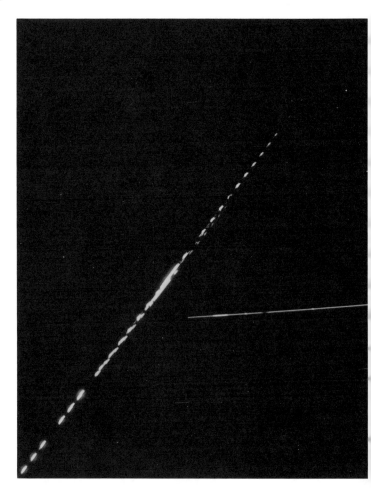

The orbit of the asteroid 1983 TB, which moves in the orbit of the Geminid meteors, entirely within the orbit of Jupiter.

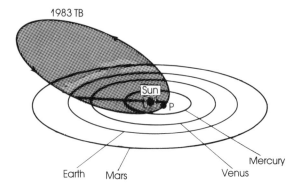

various techniques, both geologic and isotopic. One of the greatest such features, still retaining a crater configuration, is Popigai in Siberia, which is 39 ± 9 million years old and 100 kilometers in diameter. The fall occurred near the time of the Late Eocene extinctions, some 37 million years ago. Three craters, 18 to 25 kilometers in diameter, in Canada and the USSR were exploded around 97 million years ago, possibly a little earlier than a well-known extinction.

The data on craters led Carl Seyfert and Leslie Serkin of Buffalo to observe in 1979 that the ancient dates seem to *cluster* around the terminations of geological eras. At most, we can hope to find and date 30 percent of such craters, because none leave records in the oceans. Seyfert noted, therefore, that we have a 70 percent chance of missing the greatest impacts, such as that of the K/T boundary, which may have caused the major extinctions. He proposed that the clustering of the crater dates represents the impacts of a number of pieces of a small Earth-crossing asteroid that had been shattered by a collision in space. Such pieces survive some 10 million years or so before they strike the inner planets. Grieve, who has been a prominent contributor to the cratering data, seriously questions the reality of the clustering, however.

Regarding a possible rain of comets from the cloud, Sidney van den Bergh at the Dominion Astrophysical Observatory has calculated that, even if very dense, interstellar molecular clouds make huge numbers of comets, these comets will leak out over tens and hundreds of millions of years. Thus the reduced concentration would not create much of a hazard. The most recent extinction of interest occurred about 11 million years ago, when the Earth was well "below" the plane of the Milky Way, but still not out of the region of interstellar clouds. Napier and Clube suggest that, 5 to 10 million years ago, we were not very far from the Orion complex, and possibly were near the group of very bright, young stars known to astronomers as Gould's belt.

In summary, then, a number of scientists believe that the extinctions recur semiperiodically at intervals of 26 to 33 million years. They conclude that the cratering record shows about the same periodicity, even though the record is less than 30 percent complete and certainly misses some of the truly great impacts that might have been capable of causing major extinctions. Milky Way crossings by the Solar System occur with much the same period. All three periods may have a physical connection, via comet showers. But there are other possibilities, such as direct physical effects on the Earth when it is passing through an interstellar cloud.

Marc Davis and Richard A. Muller of the University of California and Piet Hut of Princeton University propose a completely different mechanism. They postulate an unseen small companion star revolving about the Sun with a period of 26 million years. This companion star, appropriately named Nemesis, would disturb the comets in the Öpik-Oort cloud, sending more or less a billion comets into the inner Solar System over an interval of about a million years at each revolution. Dozens of comets in these showers would strike the Earth and thereby cause extinctions. But, to move with a period of 26 million years, Nemesis must have an orbit with its long axis twice 87,000 AU. Before Nemesis can dip into the Öpik-Oort cloud at less than 40,000 to 60,000 AU from the Sun at each revolution, its aphelion must stretch out to about twice the radius of the Öpik-Oort cloud, more than 100,000 AU. The stability of such a long orbit seems highly uncertain in view of the attraction of passing interstellar clouds. If Napier and Clube are correct, Nemesis would be swept away in a few revolutions, if not by its first passage near an interstellar cloud. Thus the Nemesis theory and the Napier-Clube theory are mutually exclusive. If one is correct, the other is wrong. Both could well be wrong. Furthermore, Nemesis would probably have decimated the Öpik-Oort Cloud by now if it evolved as a companion to the Sun some 4.6 billion years ago.

Grieve points out enough is known about a few of the impact craters to conclude that they were produced by meteorites, and, therefore, by asteroids, not comets. A prime example is the tiny—slightly more than a kilometer in diameter—meteor crater in Arizona. It was splashed out by an iron meteorite. As modern analytic techniques grow more refined and are applied more extensively, both to ancient impact craters and possibly to the geologic strata at the levels of the extinctions, the nature of the impacting bodies should become better known. Such knowledge may not enable us to distinguish comets from stony asteroids, however. Öpik long ago suggested that the inner cores of very large comets may be strong rocky masses and that some of the Earth-crossing asteroids may be old comet nuclei. Napier and Clube hold strongly to this idea, and, indeed, suggest that the supply of Earth-crossing asteroids would long since have been decimated, had they not been replaced by the cores of large, tamed comets. Their examples are P/Encke, the recently discovered asteroid 1983 TB, which moves in an orbit like the Geminid meteors of December, and a few other asteroids that move in orbits that are highly inclined to the planetary plane, more like the orbits of tamed comets than of "main-belt" asteroids.

With regard to the great extinctions, my bet would be at least even money favoring each of the following wagers: *Some* of the major extinctions were caused by great impacts; neither great interstellar molecular clouds nor a Nemesis companion to the Sun triggered a pseudoperiodicity in the extinctions; comets produced few impact craters on the Earth, if any, and no extinctions, except possibly for some extremely large and old comets whose cores were heated to become superficially indistinguishable from asteroids. This last opinion rests on arguments mentioned by Sekanina, discussed below, that comets entering the Earth's atmosphere would be largely broken up by the drag pressure and would not reach the surface as large intact solid bodies. Whether the causes of the major extinctions were primarily asteroid impacts, comet impacts, both, or neither, we can expect that continued research will eventually answer the question. In the meantime, we can enjoy the excitement of the speculations and follow with interest the theories and the observations as they converge on the correct explanations. But whatever the true cause of the extinctions, we must bear in mind that *asteroids and comets have and will*, strike the Earth, causing serious if not catastrophic devastation.

How serious is the danger? In 1908 a collosal fireball devastated a forest and marsh region of Siberia over an area some 30 kilometers in diameter. This event is known

Tunguska desolation.
(Photograph by L. A.
Kulik.)

as the great Tunguska explosion. At a trading post 70 kilometers from the center, witnesses first felt a scorching
heat, then were knocked over by the shock wave. A roar of
thunder followed. The shock wave traveled completely
around the world twice, as registered on many baro-
graphs measuring the atmospheric pressure. For a few
nights, the sky in Eastern Europe became too bright for
quality telescopic observing because of material in the
atmosphere that had been raised by the explosion. The
blast wave and heat killed thousands of reindeer but no
humans. R. Ganapathy, a meteoriticist of Phillipsburg,
New Jersey, has analyzed small metallic spheres found
near the Tunguska fall and microparticles that the winds
carried to the Antarctic ice cap. His study indicates that
the object was stony, weighing more than 7 million tons,
and probably had a diameter of some 200 meters. Most
scientists have attributed the Tunguska fireball to the
impact of a comet, because no sizable craters or meteorites were found. In 1983, Zdenek Sekanina at the Jet Propulsion Laboratory produced a convincing argument that
the body was really asteroidal: It penetratrated to less
than 8 kilometers altitude before exploding. He showed
that a weak cometary body would have shattered at a

much greater altitude under the great ballistic pressure induced by its final velocity of some kilometers per second. Luckily, the explosion did not occur over a large town or a small city. It would have killed many or most of the inhabitants.

In 1947, a 100-ton iron-and-nickel meteorite struck Siberia, the Sikhote-Aline fall. It shattered into hundreds of pieces at an altitude of some 8 kilometers. Numerous people saw the scintillating fireworks display, reportedly brighter than the morning Sun. A 2-ton chunk excavated a crater some 30 meters across and 6 meters deep. One hundred and twenty craters more than half a meter in diameter were found. Had it landed in a city, it would have devastated a few square blocks. Undoubtedly, some such falls have occurred in our oceans in this century, but have gone unreported.

In 1972 the August tourists in the Rocky Mountains were treated to a brilliant afternoon display: a fireball traveling north, followed by a glowing rain. It missed Montana by some 55 kilometers and arched away into space beyond Edmonton, Canada, just flicking through the Earth's high and rare atmosphere. Had the object struck the Earth, it probably would have caused much more damage than the 1947 Siberian fall.

Thus, in the twentieth century, the Earth has been peppered by two rather violent impacts and one well-documented near-miss. Grieve lists nine impact craters comparable to the Arizona crater, or larger, that have formed on land in the last 2.5 million years. This adds up to about one such impact somewhere on Earth every few tens of thosands of years—definitely a minimum figure, because some craters may not yet have been identified. The Arizona crater was formed about 25 thousand years ago. Lesser impacts, such as Tunguska, could do inestimable damage. So far, all such impacts are completely unpredictable. One could happen tomorrow.

The chance that a comet can collide with the Earth is best calculated from the statistics on comets that have come near the Earth. The following table lists the ten comets that have passed nearest to the Earth with the minimum distances in millions of kilometers and in Earth radii, as derived by Zdenek Sekanina and Donald K. Yeomans.

	MISS DISTANCE	
COMET	Million Kilometers	Earth Radii
1491	1.41	220
1770 (P/Lexell)	2.26	354
1366 (P/Tempel-Tuttle)	3.43	537
1983d (IRAS-Araki-Alcock)	4.68	734
837 (P/Halley)	5.00	783
1800 I (P/Biela)	5.48	859
1743 I	5.83	915
1927 VII (P/Pons-Winnecke)	5.89	924
1014	6.09	955
1702	6.54	1025

Note: The distances for comet 1491 may be significantly in erro
The Moon's distance is 0.384 million kilometers or 60 Earth radii.

Noteworthy is the fact that five of the ten comets in the table are periodic. Two others, 1743 I and 1702, may also be periodic. The orbits of the latter two are inclined less than 5 degrees to the Earth's orbit and their periods are completely indeterminate.

Sekanina and Yeomans calculate that the chance of a comet colliding with the Earth is once in about 40 million years. This conclusion, of course, assumes that the sample represents the comet apparition rate over a time span some 100,000 times the span of good observations.

Can we do anything to prevent such "heavenly" disasters? If so, should we? The answer to the first question is: Definitely yes! We have the power to prevent small asteroids and tame comets from striking the Earth. Our conquest of space permits us, if we choose, to detect and track such bodies in their orbits. If one seemed to be near a collision course with the Earth, we could put reflectors on it, or a transponder (a radio receiver that detects and transmits radar pulses), to give us a precise orbit by radar from Earth. If the object were actually on a collision course, we could divert it. The diversion tactics would depend upon the time schedule of the collision.

With several years' notice, we could install a jet propulsion system on the body to change its orbit. If the collision date were found to be only a few months away, we might be compelled to use more drastic measures, perhaps a properly placed nuclear bomb. In this case, the problem would be to divert the body without blowing it to pieces; otherwise some of the pieces might strike the Earth, with unhappy consequences! A wild comet, rather than a tame one or an asteroid, could, however, upset our best-laid plans. We might have difficulty in obtaining sufficient advance warning. Nevertheless, we could reduce the chances of a serious impact to once in millions of years and completely eliminate catastrophic extinctions of life on Earth by impacts. Protection of the Earth from undesirable impacting bodies is not just a science fiction project for some improbable future. The cost might be comparable to, even smaller than, the world's current military expenditures. We could choose to do it now. We could choose to protect ourselves from asteroids and comets rather than from each other.

Chapter 24   Space Missions to Comets

The Space Age has already brought us new insight into the nature of comets. The future stretches before us with unlimited horizons—depending on how we choose to use these magical resources. A number of cometary missions are actually in preparation and plans for more are under way. The momentum is up for finding out more about these primitive denizens of the cosmic jungle.

Five robot spacecraft are planned to hurtle by Halley's comet in March 1986, each armed with an array of sophisticated scientific instruments to scrutinize the comet in as many ways as limitations of weight, electric power, and telemetering (radio transmission) prowess will permit. Because of financial limitations, the United States sends no mission, although American scientists and engineers made plans and recommended missions during the 1970s, well enough in advance for implementation. The Japanese, surprisingly, are sending two spacecraft on their country's first interplanetary venture beyond Earth-orbiting artificial satellites. The Soviet Union plans to send two probes carrying instruments from eight countries including France and West Germany. The Giotto mission, loaded with instruments designed and built by scientists and engineers from several countries, is sponsored by the European Space Agency (ESA).

The first probe that is expected to encounter Halley's comet is the USSR's VEGA 1, launched December 15, 1984, and headed not for Halley's comet but for the planet Venus—followed by its identical twin, VEGA 2, launched two days later. These spacecraft may appear to have been named for the brightest star in the constellation of Lyra, but VEGA is actually a Russian acronym for Venus and Halley. As their names imply, the twin probes represent a double mission. In June 1985, as they go by Venus, each is scheduled to drop a balloon the size of a basketball into the atmosphere of the planet and a Venera-type probe to land softly on the night side. The maneuver at Venus serves a subtle purpose, to utilize the planet's gravity as a

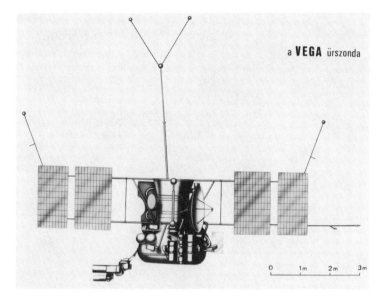

a **VEGA** ürszonda

Soviet VEGA spacecraft. Two spacecraft were launched in 1984, and after deploying Venus landers, they will fly by comet Halley in March 1986. (Photo by USSR Academy of Sciences; courtesy NASA/Jet Propulsion Laboratory.)

slingshot to bounce the VEGAs into orbits so that the spacecraft will intercept Halley's comet at a speed of 78 kilometers per second (175,000 miles per hour) on March 6 and 9, 1986. If all goes as planned, the VEGAs will pass about 10,000 kilometers (6,000 miles) on the sunward side of the comet, well inside the coma. Because the spacecraft will strike the dust grains near the comet with meteoric speed, each grain will act as a hypervelocity projectile, capable of penetrating a metal skin ten times the diameter of the grain. The risk to the spacecraft and their delicate sensing instruments near comet Halley is clearly of concern to all.

The Japanese hope to avoid the meteor missile hazard by missing the comet by a very much larger margin. Their MS-T5 is essentially a prototype to test out the spaceworthiness of their Planet-A space probe. If successful, it will pass the comet some millions of kilometers deep in the tail of Halley's comet on March 8, 1986. Its launch on January 7, 1985, preceded the August launch of Planet-A in order to test Japan's new venture into deep space. If the new launch vehicle and the probe meet their specifications in practice, Planet-A will pass the comet on the sunward side at the safe distance of 200,000 kilometers (124,000 miles), also on March 8, near the dates of the VEGA and Giotto missions.

Giotto, to be launched in July 1985, is the daredevil of the five probes to Halley's comet. Its designated trajectory misses the nucleus by only 50 kilometers (300 miles)—the uncertainty factor is about 250 kilometers—on the sunny side, only five days after Planet-A's encounter. In this region of P/Halley's coma, the dust is so hazardous—it is moving at a speed of 69 kilometers per second (154,000 miles per hour)—that Giotto's survival for the entire mission is seriously in jeopardy. A simple calculation involving the velocities and distances of the missions, as outlined above, makes the hazard more than ten times greater for Giotto than for the VEGAs. Giotto and the VEGAs are protected against the cometary missiles by improved versions of the "meteor bumper" I invented in 1946. A thin skin of metal placed a few inches outside the main shell of the spacecraft, or around delicate instruments, explodes the dust missiles on impact, so that only vapor strikes the main shell or instruments. This device greatly reduces the damage from impact or penetration. Giotto employs a double meteor bumper and the VEGAs a multiple aluminum shield.

Dust will surely cripple some equipment on both the Giotto and VEGA missions because not all the surfaces can be well protected—particularly the solar cells, the first optical surfaces, and some other surfaces that directly measure cometary gas or radiations. Judicious orientation of the sensitive elements, however, can reduce some of the hazard. The designers can only hope that dust balls the size of grapes will not strike critical instruments, communication systems, or control elements, at least until after the results of the experiments have been telemetered safely to Earth.

All the missions measure some aspects of the solar wind. The Japanese craft concentrate solely on the plasma, that is, on magnetic fields, ions, and electrons.

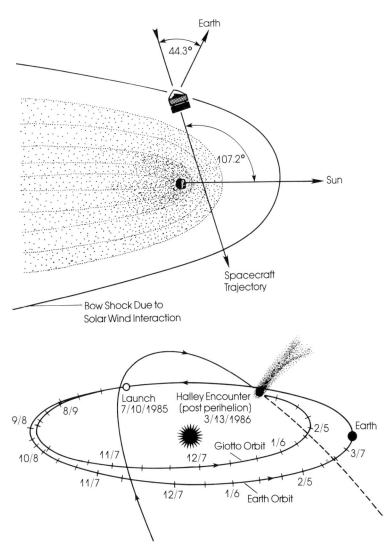

Giotto's encounter with Halley's comet, not drawn to scale. The minimum distance is planned for 500 kilometers, but the bow-shock wave with the solar wind may be 30,000 kilometers or more distant from the nucleus.

Earth

44.3°

107.2°

Sun

Spacecraft Trajectory

Bow Shock Due to Solar Wind Interaction

The orbit of the Giotto spacecraft from launch on July 10, 1985, at encounter with Halley's comet on March 13, 1986.

Launch 7/10/1985

Halley Encounter (post perihelion) 3/13/1986

9/8

8/9

10/8

11/7

11/7

12/7

12/7

1/6

Giotto Orbit 1/6

2/5

2/5

Earth

3/7

Earth Orbit

They also image the hydrogen cloud around the comet in ultraviolet light. Thus the Japanese missions complement the other missions in providing an overview of the extended activity of the comet and the solar weather in its neighbohood. The other missions will also monitor the solar wind.

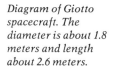

Diagram of Giotto spacecraft. The diameter is about 1.8 meters and length about 2.6 meters.

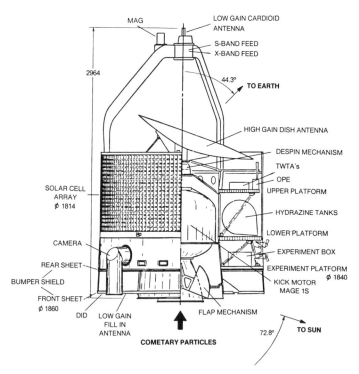

My interest, understandably, is biased toward the goal shared by the hazardous Giotto and VEGA missions, which is to obtain direct pictures of the nucleus. After more than three decades of study and speculation, I still cannot really be confident about the physical appearance of a comet nucleus. Is it spotted? Is it really grotesque in shape? The answers can come only from space missions.

The chances of imaging the nucleus, however, depend on two critical factors other than the general problems of deep-space missions: how dificult it will be to center the imaging systems on the nucleus, and whether the nucleus is hiding within an envelope of dust. Were Halley's comet less exuberant, the nucleus should certainly be fully exposed to "eyes" of the spacecraft. The visibility of the nucleus, as calculated by a number of scientists, depends entirely on the model chosen for calculating the dust emission. Perhaps the nucleus may peek out from time to time.

The VEGAs and Giotto have new and different television systems, both untried in space, and both acting as self-controlled robots to point at the nucleus during the few minutes that close-up pictures can be taken. Note that the spacecraft travel 40,000 kilometers in ten minutes or less and swing by the nucleus of the comet so fast that

The Japanese Planet-A spacecraft. This probe will encounter comet Halley on March 7, 1986, and will measure the reflectance of cometary particles in ultraviolet light. (Photo by Japanese Science and Technology Agency; courtesy NASA/Jet Propulsion Laboratory.)

ground control is impossible. At a distance of about 1 AU from the Earth, the radio round-trip time from the spacecraft is some seventeen minutes!

The pointing platforms on the VEGA carry three sensors, a short- and a long-focus television camera, and a spectrometer sensitive to visible, infrared, and ultraviolet light. As Giotto spins around every four seconds, the slits of the sensors swing by the nucleus and build up images in various colors that can be changed by filters on successive revolutions. The images are stored electronically and can be telemetered back to Earth on schedule or demand. The spin-axis of Giotto points away from the nucleus, and a controlled mirror reflects the comet's light through an optical system onto the slits. For both Giotto and the VEGAs, a robot brain chooses the bright center of the coma as the nuclear region to be imaged.

Sophisticated, lightweight instruments in Giotto measure the gas composition, cometary ions, solar-wind ions, electrons, magnetic fields, dust impacts, and composition of the dust. The VEGAs carry somewhat similar self-contained physical laboratories for measuring the physical and chemical properties of the inner cometary coma. To target the nucleus of Halley's comet within 500 kilometers, Giotto must be able to track the comet through space with an accuracy never before achieved in orbit calculations. Better accuracy than this, however, is commonplace for tracking space probes. Thus, ground-based astronomers must follow the comet diligently with position measurements and calculations to allow for the unknown variations in the motion that the comet itself induces by its jet action. In 1983, VEGA officials agreed to update their observations for use by the European Space Agency. As a result, late corrections to the Giotto motion can ensure an accurate trajectory through the inner coma. Detailed en route observations of the comet's activity can also optimize the effectiveness of some of the experiments.

In support of the space missions and direct research on Halley's comet, NASA will fly a Space Shuttle payload, the Astro-1, in early March 1986. The satellite will carry ultraviolet sensors and visual wavelength cameras. Other spacecraft in orbit, such as the Solar Maximum Mission

Ray Newburn.

spacecraft and the Pioneer-Venus Orbiter, may assist in monitoring Halley's comet. The Orbiter can image the comet at moderate range in the far-ultraviolet, particularly in hydrogen and oxygen radiation from December 1985 to March 1986.

To expedite and encourage observations of Halley's comet from the ground, the International Halley Watch (IHW) provides international centers for information and also committees to advise and help both professional and amateur observers make the best use of their efforts and telescopes in studying the comet. The prestigious International Astronomical Union at its 1982 meeting in Patras, Greece, endorsed the IHW as its official international coordinating agency for Halley's comet research. The IHW is headed by Jürgen Rahe, and its prime mover is Ray Newburn at NASA's Jet Propulsion Laboratory in Pasadena, California. In Europe, the headquarters are at the Remey Observatory of the University of Erlangen-Nurenberg, West Germany. By June 1984, 875 scientists

Insignia of International Halley Watch.

from 47 countries had joined the IHW. The goals of the IHW are to maximize the effectiveness of ground-based observations of Halley's comet by mutual worldwide communication and to provide centers for data storage and retrieval. Scientists who wish to study P/Halley can plan their programs with full knowledge of other research, so as to avoid overlap and to fill lacunae in the observations. Later they can ascertain what information has been gathered and how to find and utilize it.

As P/Drommelin made its fifth observed perihelion passage in its twenty-eight-year orbit in 1984, it was used to test IHW's communication and coordination system, which was found to have both strengths and weaknesses. The test also provided new data about this previously little-observed comet.

A powerful space tool that will be used sparingly to study comets is the Hubble Space Telescope. Its 2.4-meter (94.5-inch) mirror, unimpeded by the Earth's tremulous atmosphere, may possibly be able to image Halley's comet throughout its orbit. If launched in time, the Space Telescope could add an enormous amount of scientific information about Halley's comet during its 1986 apparition, particularly about the jets and chemistry near the nucleus. Ultraviolet capability coupled with high-resolving power makes the Space Telescope the fulfillment of an astronomer's dream for studying the universe.

Comet Halley is not the first cometary target for a scientific probe. The mission of a joint NASA–ESA venture, the International Sun Earth Explorer Three (ISEE-3) was changed to the International Cometary Explorer (ICE) in December 1983. The ISEE-3 was launched from Earth in August 1978. It measured properties of the solar wind and the Earth's magnetosphere as it traveled well beyond the Moon toward the Sun. By a slingshot maneuver in October 1982, the Moon threw ISEE-3 into an orbit to explore the solar- (down-) wind region of the geomagnetic tail. Again, in December 1983, the Moon became the slingshot to put ISEE-3 into an orbit aimed to fly the spacecraft by periodic comet Giacobini-Zinner on September 11, 1985, barely outside the Earth's orbit. The last lunar maneuver was a triumph of space navigation, as ISEE-3 had to swoop down to 120 kilometers above the Moon's surface so that the Moon's gravity could shift the artificial satellite's orbit properly.

P/Giacobini-Zinner orbits the Sun with a period of 6.5 years and barely dips within the Earth's orbit. Its chief claim to fame rests in its meteor stream, the Draconids, which in 1932 and 1936 provided two magnificent displays. The meteor bodies are true dust balls with extremely low densities, probably less than 1/20 the density of water. The orbits of the Earth and the comet so nearly intersect that the meteor bodies are relatively pristine after being blown off from the comet. In any case, they represent the extreme case of fragile solids from comets, and are typical cometary ones in that their densities are about one-third that of water.

Space probe ICE is equipped to measure magnetic fields, particles, and ions from the solar wind, particularly when the solar wind reacts with the gases from P/Giacobini-Zinner. ICE is virtually a plasma laboratory placed in the tail of the comet.

The six cometary missions in 1985 and 1986 should greatly enhance our knowledge of the abstruse physics of ion tails and the complex chemistry and physics of the coma. The resulting theoretical models should then clarify the multitudinous observations of other comets and thus help to bring their real character into focus. If the shy nucleus of Halley's comet pokes its head through its

The International Cometary Explorer (ICE) reborn from the International Sun-Earth Explorer Three (ISEE-3).

dust halo to have its picture taken, it will answer a host of exciting questions. And there is a strong possibility that we will learn much more about the exotic ices in comets, and find additional clues to their origin and, indeed, to our own origins. Previous experience in space missions, however, suggests that unexpected results are the ones most to be expected.

Whatever the outcome of these missions, they will act as a theoretical springboard for the next vital step in studying comets—a *rendezvous* mission in which a robot spacecraft flies in orbit near a comet and observes it from all aspects, for months on end, with a battery of scientific instruments. The United States via NASA is planning just such a rendezvous mission to P/Kopff in 1995. In 1983, an official Comet Rendezvous Science Working Group began its deliberation. The stated objectives for the mission centered first on all measurable properties of the nucleus, with secondary but vital interest in the physical and chemical processes in the coma and tails. Next to the ideal—sample-return mission—a rendezvous mission is the best means of studying a comet. Flyby missions, such as those to P/Halley, compress almost all the scientific

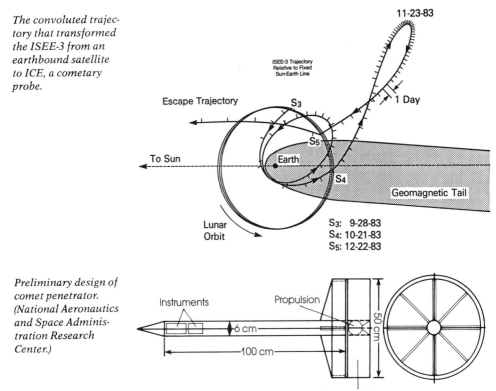

The convoluted trajectory that transformed the ISEE-3 from an earthbound satellite to ICE, a cometary probe.

Preliminary design of comet penetrator. (National Aeronautics and Space Administration Research Center.)

observation into minutes, whereas a rendezvous allows observation to stretch out over a period of months. Furthermore, the rendezvous spacecraft can maneuver within a few kilometers of the nucleus in any desired direction to permit observation of all the activity in the nucleus, coma, and tails from the best vantage points.

The mission may include a meter-long dart to spear the comet nucleus. This *penetrator* would be spun up and fired from the spacecraft when it is near the nucleus to enter the nucleus like an oven thermometer in a roast turkey. An accelerometer in the penetrator would register the shock as the point of the dart entered the dirty ice to measure its physical strength; other instruments would measure the temperature at depth and the abundances of some twenty chemical elements by their gamma rays induced by cosmic rays. All these measurements, of course, would be radioed to the spacecraft and then relayed to the investigators on Earth.

Periodic comet Kopff was chosen for the mission because it is a "reliable" comet in an appropriate orbit (its period is 6.4 years), moving at a tilt of only 4.7 degrees to the Earth's orbit, and passing perihelion at heliocentric distance of 1.47 AU. The more exciting P/Encke might have been chosen, but it would carry the mission down to 0.34 AU of the Sun, where solar heating would seriously affect several of the delicate instruments. P/Kopff is considered "reliable" because it is moderately bright among tame comets, and has demonstrated interesting activity during its ten observed apparitions since its discovery by the German astronomer August Kopff in 1906. In other words, P/Kopff appears to be a "healthy" comet, not likely to fade away before 1995. Its nongravitational motion is intriguing because its orbital period increased between 1906 and 1946, but has been decreasing since then. In comparison with P/Encke, its spin-axis motion must be very rapid, perhaps because it has an oddly shaped nucleus.

Choosing the specific instruments to be flown in such missions is a fascinating process. All participants are motivated and exhilarated by the prospects of new scientific results beyond present frontiers. The scientists all know that their proposed experiments would be groundbreaking and worth the expenditure of a decade of development. Vital participants are the budget representatives from Washington (both NASA and the National Science Foundation), and the spacecraft engineers, who carry the final responsibility for the practical success of the mission. The amount of money available for the instruments, including development, manufacturing, and testing is always limited. The total weight is set at about 100 kilograms (220 pounds) for all the experiments, controls, and information storage and transmission. Power is limited, as is the telemetering, or message transmission rate. Experiments must not interfere with each other or with spacecraft operation. Space and weight distribution must always be considered. Understandably, there is competition for the optimum location on the spacecraft.

The heated debate over objectives and detailed plans extends over several meetings of two or three days each, until finally some compromise is reached. In the end, fortunately, rational thinking largely controls the result. The losing experimenters hide their disappointment while the winners hope that they can really meet the severe cutbacks imposed in weight and power as well as the other restrictions on their respective instruments. Their next trial will be the final competition for an experiment on the spacecraft after NASA has circulated its *Announcement of Opportunity* to all who may be interested.

The P/Kopff rendezvous orbit will almost certainly fly by an asteroid and provide at least a clear picture of a cometary cousin. Whether or not the spacecraft will finally be landed on the nucleus of the comet is undecided. If so, astronauts of the future may land on P/Kopff and possibly find the spacecraft perched on the top of a hillock, well dusted over with meteoric debris. The spacecraft will probably be less dark than the cometary surface so that the landscape around it would sublimate away, leaving the spacecraft exposed on a hillock. Possibly the hillock may have wasted somewhat, so that the spacecraft might be tilted over on its side or even possibly blown entirely away into space by the sublimating ices.

Even a rendezvous mission cannot answer some of the most critical questions about a comet, particularly questions about its age and its precise composition and chemistry. No remotely controlled laboratory in a spacecraft can match the accuracy and thoroughness of a laboratory on Earth. Therefore, we must snatch a sample from a comet and bring it back to Earth for analysis. The simplest of such missions would collect dust near the nucleus in a flyby of a comet. The spacecraft returns to an obit about the Earth, a difficult maneuver. A shuttle then collects the sample and decontaminates it to protect us from bacteria or viruses. I hope that the cheaper direct reentry mode can be accepted when a sample return mission is activated because I consider the hazard to be zero. Even if comets contain bacteria or viruses, they have been landing on Earth since its infancy.

When will we set foot on a comet? Footprint of Neil A. Armstrong or Edwin E. Aldrin, Jr., on the Moon, July 24, 1969. Whereas their footprint should last for millions of years on the Moon—unless destroyed by a tourist—on a comet one would probably not outlast a single apparition. (Courtesy National Aeronautics and Space Administration.)

Epilogue

Our commentary detective story has carried us through
many centuries—from the superstitions of antiquity to
the sophisticated observations made by space vehicles.
The pace of learning has quickened during the four centuries since Tycho Brahe first proved that comets are celestial bodies. Our expectations run high. Perhaps within the
reader's lifetime someone will set foot on a comet and
provide answers raised by the studies of comets. What
further surprises are in store for us as we explore the con-
tents of these mysterious packages preserved in deep
freeze since the Solar System was young?

Fred L. Whipple

Additional Readings

Brandt, John C. *Comets.* San Francisco, Calif.: W. H. Freeman, 1981. [A compendium of relevant articles from *Scientific American.*]

Brandt, John, C., and Robert D. Chapman. *Introduction to Comets.* New York: Cambridge University Press, 1981. [For the advanced amateur or professional astronomer.]

Brown, Peter Lancaster. *Comets, Meteorites and Men.* New York: Taplinger, 1974. [Well packed with information for the serious layman.]

Calder, Nigel. *The Comet Is Coming.* London: British Broadcasting Corporation, 1980. [A cynical but frequently amusing account of Halley's comet and others.]

Chapman, Robert D., and John C. Brandt. *The Comet Book—A Guide for the Return of Halley's Comet.* Boston: Jones and Bartlett, 1984. [Elementary.]

Delsemme, Armand. *Comets, Asteroids, Meteorites: Interrelations, Evolution and Origins.* Toledo, Ohio: University of Toledo, 1977. [For the advanced amateur or the professional astronomer.]

International Astronomical Union Circulars. Cambridge, Mass. [About 130 circulars published a year to notify astronomers of new astronomical discoveries, including comets. Available from Central Bureau for Astronomical Telegrams, 60 Garden Street, Cambridge, Mass. 02138. For the serious amateur and for professional astronomers. No illustrations.]

International Comet Quarterly. Cambridge, Mass. [World's largest journal devoted solely to comets. For the advanced amateur and professional astronomers. Illustrated. Available from Daniel W. E. Green, Editor, Smithsonian Astrophysical Observatory, 60 Garden Street, Cambridge, Mass. 02138.]

International Halley Watch Amateur Observer's Manual for Scientific Comet Studies. Parts I and II. Cambridge, Mass.: Sky; Hillside, N.J.: Enslow, 1983. Also available from U.S. Government Printing Office, Washington, D.C. [Excellent observing manual for amateur astronomers. Includes star charts for viewing Halley's comet in 1985–86.]

Kronk, Gary W. *Comets: A Descriptive Catalog.* Hillside, N.J.: Enslow, 1984. [Textual descriptions of cometary apparitions from 381 B.C. through 1981. Generally a reliable reference source, although researchers should consult original sources. No illustrations.]

Marsden, Brian G. *Catalog of Cometary Orbits.* 4th ed. Cambridge, Mass.: Minor Planet Center, Smithsonian Astrophysical Observatory; Hillside, N.J.: Enslow, 1982. [Lists orbital elements, dates of observations, statistics, and names of comets. For amateur and professional astronomers. No illustrations.]

Sky and Telescope. Cambridge, Mass. [A popular astronomy magazine; has a monthly article on comets.]

The Study of Comets. 2 vols. Washington, D.C.: NASA, 1976. Available from Superintendent of Documents, U.S. Government Printing Office, Washington, D.C. 20402. [More than sixty papers covering most areas, of cometary research, for professional astronomers. Technical illustrations.]

Wilkening, Laurel L., ed. *Comets.* Tucson, Ariz.: University of Arizona Press, 1982. [For the advanced amateur or the professional astronomer.]

Index